In India, the problem of individual liberation i.
Hindu philosophy by numerous scholars, especially social reformers in the
19th and 20th centuries. Vivekananda (1863–1902) belonged to a branch of
Hindu philosophy called Vedanta (see Sriraman & Benesch, 2005), in particular to a special strand of Vedanta, which holds that no individual can be
completely free unless every one else is also free (from oppression). In other
words, we as individuals are obliged to act to better society. Vivekananda was
able to move beyond the prevalent dogmatic caste system which characterized Indian society and propose a theory of action which necessitated that
each of us consciously act towards bettering the lot of our fellow humans, if
our goal is to ultimately liberate ourselves and become enlightened.

From a Freirean perspective it is not possible to "empower people…"—
the best we can do is to create conditions to facilitate, support people empowering themselves, and to work along side in common struggle. I prefer to view the individual chapters in this book from this perspective. The
chapters can also be viewed as conscious and well directed action from the
various authors aimed at education creating real *Meaning*.

ACKNOWLEDGEMENTS

I gratefully acknowledge the help of Ubiratan D'Ambrosio in sections of
this chapter. This chapter is dedicated to my teachers and friends: Viola
Cordova (posthumously), Walter Benesch and Harry Adrian for being living examples of conscious action.

NOTES

1. *Nihilism* is a philosophy written on by Martin Heidegger as well as Friedrich
   Nietzsche. Although the writings on Nihilism of these two eminent philoso-
   phers have been subject to apposing interpretations, the basic premise of nihilism is that the world or existence as we know it is ultimately without any objective meaning, with obvious implications for organized religion, morals and
   ethics. Neitzsche's nihilistic position stems from the frustration in our search
   for meaning. In Russia, nihilism was associated with revolution that rejected
   the authority of church and state. I view the Existentialism movement of the
   20th century as an attempt to resolve the problem of meaningful existence if
   one embraced nihilism.
2. See Richard Dawkins (1964) *The Selfish Gene*
3. Asoka (c. 299–237 BCE) is credited with the establishment of the so-called
   "first" Indian empire, accomplished through decades of bloody conquests.
   His deep remorse over the carnage at Kalinga led him to embrace the peace-
   ful doctrines of Buddhism. Under his protection, Buddhism flourished and

numerous Buddhist texts were written. Asoka also sent numerous emissaries of Buddhism to places like South East Asia, Egypt, Libya, and Macedonia, which resulted in the "golden" age for Buddhism.

4. Karl Marx, Friedrich Engels, *Gesamtausgabe*, Edited by the Institut für Marxismus-Leninismus

5. *World Resources 2000–2001: People and Ecosystems: The Fraying Web of Life*. United Nations Development Programme, United Nations Environment Programme, World Bank, World Resources Institute.

6. It is unclear whether Freire was Marxist, or the book was a Marxist work. Freire was clearly influenced by Marx and Che Guevara. In the introductory section of the book Freire wrote: "I am certain that Christians and Marxists, though they may disagree with me in part or in whole, will continue reading to the end."

7. My discussion with Rico Gutstein and Ubi D'Ambrosio was very helpful in understanding Freire.

8. Program for International Student Assessment

9. Here Cho & Lewis are synthesizing the writings of Ellsworth (1989), Gore (1990) and Weiler (1991). These particular writings convey a completely different conception of the complexities of empowerment from the point of view of feminist pedagogy. See references.

REFERENCES

Cho, D., & Lewis, T. (2005). The persistent life of oppression: The unconscious, power and subjectivity. *Interchange: A Quarterly Review of Education*, 36(3), 313–329.

Comte, A. (1972). Das Drei-Stadien-Gesetz. In H.P. Dreitzel (Ed). *Sozialer Wandel* (pp. 95–111).Neuwied and Berlin.

Darwin, C. (1871). *The Descent of Man*. London: John Murray.

Dawkins, R. (1976). *The Selfish Gene*. Oxford: Oxford University Press.

Ellsworth, E. (1989). Why doesn't this feel empowering? Working through the repressive myths of critical pedagogy. *Harvard Educational Review*, 59(3), 297–324.

Freire, P. (1998). Pedagogy of freedom : ethics, democracy, and civic courage. Lanham: Rowman & Littlefield Publishers.

Gore, J. (1990). What can we do for you! What can "we" do for "you"? Struggling over empowerment in critical and feminist pedagogy. *Educational Foundations*, 4(3), 5–26.

Gutstein, E. (2006). *Reading and Writing the World with Mathematics: Toward a Pedagogy for Social Justice*. New York, Routledge.

Organization for Economic Co-Operation and Development [OECD] (2004). *Problem Solving for Tomorrow's World—First Measures of Cross Curricular Competencies from PISA 2003*, http://www.pisa.oecd.org/dataoecd/25/12/34009000.pdf. Retrieved September 29, 2005.

Marx, K., & Engels, F. (1879–1882). *Gesamtausgabe*, Edited by the Institut für Marxismus-Leninismus.

Radhakrishnan, S. (1964). *The Dignity of Man and the Brotherhood of Peoples.* Foreign Affairs Record of the Government of India, (January, 1964).

Sriraman,B., & Törner, G. (2007). Political Union/ Mathematics Education Disunion: Building Bridges in European Didactic Traditions. (In press) in L. English (Editor). *Handbook of International Research in Mathematics Education* (2nd ed.). Mahwah, NJ: Erlbaums.

Sriraman, B., & Benesch, W. (2005). Consciousness and Science: An Advaita-Vedantic perspective on the theology-science dialogue. *Theology and Science,* 3(1), 39–54.

Weiler, K. (1991). Freire and a feminist pedagogy of difference. *Harvard Educational Review,* 61(4), 449–474.

CHAPTER 2

HOME, SCHOOL, AND COMMUNITY PARTNERSHIPS IN NUMERACY EDUCATION

An Australian Perspective

Merrilyn Goos
The University of Queensland, Australia

Tom Lowrie
Charles Sturt University, Australia

Lesley Jolly
The University of Queensland, Australia

ABSTRACT

The importance of building educational partnerships between families, schools and communities is increasingly acknowledged since family and community involvement in education is thought to be associated with children's success at school. Yet there are discrepancies between the rhetoric of policy documents and the practice of family and community involvement in

International Perspectives on Social Justice in Mathematics Education, pages 11–35

education. This paper draws on a large Australian study to critically examine different perspectives on numeracy education partnerships, with particular emphasis on the extent to which the needs of educationally disadvantaged children were being met. We elaborate a framework for analysing key features of educational partnerships, and then use the framework to compare the features of effective numeracy education partnerships represented in two case studies from our study. The case studies highlight different ways of initiating partnerships, different perspectives of stakeholders, different numeracy practices, and different ways of responding to cultural diversity and geographical isolation.

INTRODUCTION

This chapter explores issues arising from research on educational partnerships between families, schools and communities in contexts where diversity and disadvantage impact on children's numeracy learning and achievement. It is widely recognised that parents and families are the primary educators of children and are responsible for laying down the social and intellectual foundations for their learning and development. This assertion is also grounded in the education research literature, conveying the clear message that parental and community support benefits children's learning, including their numeracy development (Cairney 2000; Epstein, 2001; Horne, 1998).

Numeracy education has become a high priority in Australia, and the government policies and strategies formulated to address this area typically capitalise on the need to build partnerships with homes and communities (e.g., Department of Education, Training and Youth Affairs, 2000). This position on partnerships is consistent with the description of numeracy proposed by Australian mathematics educators: "to be numerate is to use mathematics effectively to meet the general demands of life at *home*, in paid *work*, and for participation in *community and civic life*" (Department of Employment, Education, Training and Youth Affairs, 1997, p. 15, emphasis added). Such an approach to numeracy implies that it is the responsibility of all members of society—schools, families and communities—to ensure that children gain not only mathematical knowledge and skills, but also a repertoire of problem solving and decision-making strategies needed for intelligent citizenship in a rapidly changing world.

Yet there are discrepancies between the rhetoric of policy documents and the practice of family and community involvement in education, as current partnership models disregard how families' material and cultural conditions and feelings about schooling differ across social groups (deCarvalho, 2001). These were some of the issues we addressed in a national research project that investigated home, school and community partnerships in children's numeracy education. We analysed features of effective partnerships in the elementary school and pre-school years, with particular emphasis on the extent to which the needs of educationally disadvantaged children were being met. In this paper we draw on two of our case studies to discuss characteristics of successful numeracy education partnerships for Indigenous communities and rural families in regional and remote parts of Australia.

EXAMINING THE CONCEPT OF "PARTNERSHIPS"

Epstein (1995) defines home, school and community partnerships as exemplifying a relationship between "three major contexts in which students live and grow" (p. 702) and in which shared interests in and responsibilities for children are recognised. In addition, Funkhouser and Gonzales (1997) state that successful partnerships involve the sustained mutual collaboration, support and participation of school staffs and families at home and at school, in activities and efforts that have a positive effect on the academic success of children in school. Because home, school and community represent the major overlapping spheres of influence in children's education and development, researchers and practitioners call for their collaboration as partners who "work together to create better programs and opportunities for students" (Epstein, 1995, p. 701).

The Role of "Home" in Home-School Relations

While recent shifts in educational policies are partly based on the recognition that good relationships between parents and schools benefit students, consensus has not been reached about how these effective relationships should be achieved, who holds responsibility for what, and where power and control should reside in making educational decisions. Despite the frequency with which the concept of "partnership" is employed, its manifestation in practice often differs from the rhetoric of educational initiatives. Cutler's (2000) historical study of connections between home and school in American education demonstrates that recognition of parental influence in children's education in practice has been often blended with the construction of parents as adversaries who are either uninvolved and irresponsible or overly demanding and intrusive. This idea echoes with Sarason's (1995) view that the present governance structures of schools define, and indeed limit, the nature and scope of parental involvement. Parents are usually invited by schools only when it is needed, and staff of some schools want parents to be involved only in specific ways and at times determined by the staff. In particular, low-income parents often feel and are treated as "less" than the professionals in schools (Fine, 1993). In relation to mathematics education in the USA, Peressini (1998) found that accepted roles for parents were constructed as ranging from spectator to partner and from the deterrent to catalyst of mathematics education reforms.

Mismatches between home and school environments and failure to recognise parental diversity can create barriers to partnerships (Crozier, 2000). Also, because numeracy events embedded in the everyday activities of families or communities (such as budgeting, shopping, scheduling, playing games, measuring or building or designing things) are less visible than numeracy events taking place in school mathematics classrooms, the school can conceive of the home as a subservient context in which the numeracy concepts and skills taught in school are to be practised and reinforced. The emergence of family numeracy programs has gone some way towards connecting home and school practices by involving parents and children together in meaningful mathematical activities (Horne, 1998). However, the various stakeholders in children's education may still have divergent perspectives on what constitutes partnerships and what their roles should be.

The Role of "Community" in School-Community Relations

Socio-cultural researchers define "community" as a "community of practice"—that is, a group of people engaged in an activity driven by common or closely intersecting goals and interests (Wenger, 1998; Wenger, McDermott & Snyder, 2002). In pursuit of these goals and interests, they employ common practices, work with the same tools or resources and use specific discourse. Communities constitute social contexts and meanings for learning as people participate in social practices. Knowledge is integrated in the doing, social relations and expertise of these communities. Furthermore, the processes of learning and membership in a community of practice are inseparable. Because learning is intertwined with community membership, it is what lets people belong to and adjust their status in the group. As participants change, their learning and their identity—relationship to and within the group—also change. Therefore, communities constitute the most powerful learning environments for children, creating potential for their development as children engage in social practices with others.

This approach to learning suggests that teachers need to understand their students' communities of practice and acknowledge the learning students do in such communities (Saxe, 2002; Sfard, 2002). Drawing on communities' funds of knowledge can capitalise on cultural diversity and overcome any mismatch between students' home environments and the culture of school. McIntyre, Rosebery and Gonzalez (2001) argue that minority and poor children can succeed in school if classroom practices give them the same advantage that middle class children have—instruction that puts knowledge of their communities and experiences at the heart of their learning. In the view of these researchers, learning mathematics is more than structured individualised cognition; it is also dependent on the social and cultural situation and values of the learner.

Community partnerships focusing on numeracy issues do not usually do so exclusively, and Hexter (1990) points out that community-based programs deemed exemplary for their interventions in support of educational access are often based on more than numeracy. Such partnerships usually take a more holistic approach, as in the instance cited by Goodluck, Lockard and Yazzie (2000) of a bilingual Native American school district in Arizona: the conceptual framework is centred more globally on issues of difference and disadvantage. In Australia, Stanton (1994) identified conflicts between "white" school practices and Indigenous values, and described a community-based and community-focused program with a curriculum organised around the symbolic, societal, and cultural components of culturally sensitive mathematics. As Kahne (1999) points out, the most important aspect of community programs is the development of long-term relationships in support of positive social change.

RESEARCH DESIGN AND METHODOLOGY

The design of our research project consisted of three phases. Thr first phase began with a questionnaire survey of education organisations and parent and community groups throughout Australia, complemented by a national email survey of elementary school principals, to obtain information on the distribution and scope of current programs and practices. A total of 499 surveys was returned and analysed. The second phase comprised interviews with key personnel in the central offices of the government and non-government education sectors in each Australian State and Territory. This allowed us to identify 38 programs or initiatives that connected schools with families and communities to support children's numeracy learning. From these we selected seven exemplary, sustained numeracy education programs that were the subject of detailed case studies in the third phase of the research. Two researchers collected data from each case study site over a period of three to six days. Methods included: observation of classrooms, school staffrooms, teacher-parent interactions, and families in their homes; interviews with teachers, school administrators, support staff, and parents; and analysis of teaching materials, policy documents, and evaluation reports.

Our framework for selecting and analysing the case studies was developed from the method we used to record and categorise key features of partnership programs identified from the interview phase of the study. The analytical framework takes into account:

- different ways of initiating and funding partnerships and their implications for parental and community involvement in numeracy education;
- stakeholder perspectives on the links between schools, families and communities;
- attention given to the needs of educationally disadvantaged children;
- the nature of numeracy practices.

The first dimension of the case study analytical framework classifies relations between educational systems, schools, families and communities in terms of how partnerships are *initiated and funded*. *Top-down* partnerships are initiated and sponsored by an education system with uniform program goals and processes across schools. *Top-supported* partnerships rely on an education system for some overall sponsorship or coordination, but schools design and control the program. *School-generated* partnerships are initiated by a school independently of an education system, although this may involve resources available from the system. *Home or community-generated* partnerships have their origins in these sectors and are designed and implemented with input from families and community members. Clearly, partnership initiation strategies and funding regimes are bound up with issues of power and authority in stakeholder relations.

The second dimension of the framework recognises the different *perspectives of stakeholders* on what constitutes partnerships and what their roles might be. We classified these as school-centred, family-centred, or community-centred. For *school-centred* perspectives we drew on Epstein's (1995) work on home-school partnerships to describe six ways in which schools understand the roles of families and communities: parenting, communicating, volunteering, learning at home, decision-making, and collaborating with the community. Less attention has been given to the ways in which families and communities might understand their connections with schools and with each other, and this in itself is suggestive of power relationships between these groups. We drew on available literature in this field (James, Jurich & Estes, 2001; Jordan, Ozorco, & Averett, 2001; Katz, 2000; Keith, 1999) to identify a range of family-centred and community-centred perspectives on partnerships and roles. The following *family-centred perspectives* describe how families might see their connections with schools and their communities:

- creating supportive learning environments at home;
- providing parental support for the child and articulating parental aspirations for the child's education;
- promoting parents as role models for the value of education;
- recognising home practices that support numeracy development;
- organising parent-directed activities that connect children to out-of-school learning opportunities;
- initiating parent-child discussions and interactions about school related issues and activities.

We propose that the following *community-centred perspectives* describe how communities might see their links with schools and families:

- initiating community-driven school reform and curricular enrichment efforts that seek to improve local schools;
- developing school-business partnerships;
- developing school-university partnerships;
- offering community service learning programs;
- offering after school programs;
- offering more extended programs that target children's and family numeracy (e.g., the Family Maths program).

The third dimension of the framework looks at ways of responding to diversity and educational disadvantage by identifying the *groups of students targeted* by the program. These include students from Indigenous (i.e., Aboriginal or Torres Strait Islander), non-English speaking, and low socio-economic backgrounds; students in geographically isolated locations; and low achieving students deemed to be at risk of failing to meet State mandated benchmarks for numeracy performance.

The fourth dimension of the framework identifies numeracy practices in each case study. We operationalised the description of numeracy quoted in the introduction to this paper by looking for evidence of three aspects of numerate practice and the type of knowledge associated with each (Willis, 1998). To "use mathematics," students need to have *mathematical knowledge* of concepts and skills. Using mathematics "effectively" requires that students have *strategic knowledge* to enable them to choose and apply mathematical concepts and skills that are appropriate for dealing with unfamiliar problems. Using mathematics effectively "to meet the general demands of life" reminds us that numeracy is context-specific because mathematics is embedded in everyday situations. Thus numerate practice requires *contextual knowledge*, and school mathematics needs to be aligned to the kind of authentic problem solving situations that individuals regularly encounter in their lives (Boaler, 1993). Because each Australian State and Territory develops its own mathematics syllabus, we drew on the source document that influenced syllabus development across Australia, *Mathematics—A Curriculum Profile for Australian Schools* (Curriculum Corporation, 1994), to elaborate on the three aspects of numeracy outlined above. This document emphasises not only mathematical content (organised under the headings of *Number, Measurement, Space, Chance and Data*, and *Algebra*), but also the development of strategic and contextual knowledge and the variety of real life contexts in which students may choose and apply mathematics (elaborated under the heading of *Working Mathematically*).

We selected the final suite of seven case studies to sample a range of partnership initiation strategies, stakeholder perspectives, and target groups of students. In the remainder of the paper we analyse two of these cases, one an innovative approach to improving Indigenous children's access to preschool education in remote communities, and the other a long standing program that provides distance education to children of rural families.

THE MOBILE PRE-SCHOOL PILOT PROGRAM

The Mobile Pre-school Pilot Program develops pre-school programs and materials to distribute to Indigenous children aged 3–5 years in remote locations in Australia's Northern Territory. Previously there was no access to pre-school education because of the small numbers of children in each community. (The government's funding formula for staffing schools required enrolment of at least 12 children in any one centre in order for a qualified teacher to be employed. However, most remote communities are too small to satisfy this requirement.) This is an example of a top-supported partnership in that it is government funded but without the requirement for uniform implementation across all sites in the Northern Territory. Our investigation of the history of the program also revealed that many elements were originally, and continue to be, community-generated, thus increasing family and community participation in making educational decisions. Although the partnership does feature school-centred perspectives on the roles of families and communities, it derives its strength from community-centred perspectives, especially the role of local communities in deciding whether and on what terms to accept the program and in gaining financial and social benefits from their participation.

The aim of the program is to increase enrolment, attendance, and participation of Indigenous children in remote areas and prepare them for formal schooling through pre-literacy and pre-numeracy activities. Materials consist of a variety of play activities and items such as painting materials, puzzles, counting, colour and shape matching games, picture story books, play dough and block construction as well as larger equipment like tricycles, prams and dolls, climbing and sand play equipment, packed into large (90 cm × 50 cm) plastic containers. Materials are developed and organised by trained early childhood teachers who prepare and store the materials in their home bases and transport them to surrounding areas by light aircraft or off-road vehicles. The play-packs are often compiled around themes such as transport, communication, colours, and insects, and are rotated between sites weekly or fortnightly, depending on the contingencies of visiting the site.

Teachers travel with the play-pack and introduce the materials to the local teaching support officer (TSO). The TSO in most cases is an Indigenous person chosen by their community to take on the role of organising and running the pre-school sessions in their area. When teachers visit individual sites they introduce the materials in the play-pack to the TSO, explaining how each item might be used. The TSO bases his or her work in the ensuing week or fortnight on the new activities provided in the current play-pack. Teachers circulate between locations, which are grouped into clusters for organisational and planning purposes. This paper deals with our case study observations in the Arnhem and Katherine regions. In the Arnhem Cluster we visited the communities of Yirrkala and Dhalinybuy, and in the Katherine Cluster we observed operations with the Bulman Indigenous community.

Observations of the Partnership in Operation

Pre-school is usually held in the morning three to five days per week, for about two hours, with a morning tea break half-way through. TSOs lead sessions with the aid of parents who follow the TSO's lead in helping the children to use the materials. Older siblings may also attend and help, and younger siblings, if present, take part in the pre-literacy and pre-numeracy activities. Food for morning tea is provided by the teacher on her visiting days and shared with others from around the community.

In Yirrkala the pre-school is run in combination with the child care centre on their premises. This was an organisation of convenience as the child care centre had lost numbers, and the principal of the local school, when faced with a similar issue, had decided to relinquish his pre-school teacher. When the Mobile Pre-school was established, it joined forces with the child care centre for their mutual benefit. Dhalinybuy has a one-teacher school in which classes are taught by a qualified Indigenous teacher. The Dhalinybuy pre-school was conducted outdoors on a large woven mat under a shady tree. The Bulman pre-school worked in conjunction with the primary school and used one of its rooms, and an open covered area.

At all the sessions we were able to observe, the visiting teacher was present and set the agenda, with support and help from the local TSO. Where the pre-school was closely associated with a school, the teacher there also played a significant role. Parents were present in a fairly liminal fashion but community support was clearly crucial, especially in deciding whether the program was to operate in their community or not. For instance, when the TSO at Dhalin-buy, who is also chair of the local school council, appealed to his community for someone to take on the TSO role he was told "No, you be the teacher." The clear implication was that community people senior to the TSO made this decision. Overall, the visiting teachers are cast in the role of experts in the field who make suggestions to the local TSOs and encourage them to adapt the program for the week to immediate circumstances. Without direct observation of days when the visiting teacher is not there it is impossible for us to say how roles are negotiated in that event. Good personal relationships between the visiting teachers and the community members seem to allow for a certain equity in the partnership, thus reinforcing the trust between partici-pants that seems to be crucial for the success of the program.

There is a dynamic interchange of activity and communication among people at all levels of local community and in the organisation of the pre-school program. The visiting teachers are very familiar with the Northern Territory and have known the people in the communities and in the educa-tional and child care organisations for many years. Communication occurs predominantly by word of mouth and the play-pack is a means of providing materials that people in local areas may use and adapt in their own ways. In the Arnhem Cluster, the teacher relies on the TSO and other parents for translation between Yolngu (the language of the local Aboriginal people) and English. In the Katherine cluster, more English is spoken, though Kriol is the home language. Since the TSOs and even the teachers are intimately involved in everyday affairs in small communities there is a transparency be-tween the program and the community that facilitates communication about routine details and individuals. Communication among teachers and TSOs is maintained not only by weekly (or at least regular) personal visits but also by bi-semester or bi-term workshops in the central location (Yirrkala or Kather-ine) where the program is assessed and future plans are made. Participants at this level can also telephone each other regularly. This is successful only because the main actors all share a long history of commitment to early child-hood education and the welfare of the communities concerned.

Numeracy Practices

The pre-numeracy activities we observed are typical of those conducted in mainstream Australian pre-schools, and aimed to develop number, measurement, space, and chance and data concepts (Curriculum Corporation, 1994). Active play with puzzles and toys such as cars required shape and colour matching as well as sequencing and counting. Songs and stories provided reinforcement of the language used to make comparison, describe size, shape and sequence and discuss ideas about chance and uncertainty. Games such as "Follow the Leader" addressed sequencing, following instructions and counting. Neither strategic nor contextual numeracy knowledge were a specific focus of this program. In fact, it is tempting to argue that many of the toys and activities provided may not have been meaningful for children whose everyday experience was living on Aboriginal homelands. Several puzzles made use of cars, trucks, traffic lights and all the accoutrements of city-based transport. It is not that the children are totally unfamiliar with such things, but there is not a close fit between red double-decker buses represented in the puzzles and the minivan that serves as a bus in their local community. Of course this lack of "relevance" would also be an issue for other Australian children whose life experiences are not represented by the play activities and materials provided to them. Nevertheless, as we discuss below, local people insisted that children needed to become familiar with the world beyond their own communities.

Context and History of the Partnership

Indigenous education in Australia has been complicated by the history of colonisation. Many studies have documented the damaging effects of attempts to transplant an education system that embodies quite different epistemologies, attitudes and normal behaviours into Aboriginal communities (Folds, 1987). In recent years "two-way education," taking something from both western and Indigenous culture, and adapting it to local conditions and aspirations, has become a popular catch-cry (Harris, 1990; Malcolm, 1999). However the pressures of mainstream culture are hard to resist, and the task is further complicated by the fact that every Aboriginal area, indeed every community, has its own history and hence its own needs and aspirations. The sites we looked at here are informative in this regard.

The Yolngu of Arnhem Land were among the last Aboriginal groups to have been directly affected by colonisation. These were largely confined to the operations of a small Christian Mission until, in the late 1960s, bauxite began to be mined on the Gove Peninsula. This was sanctioned by government without consultation with local Yolngu landowners. Legal disputes over this issue led ultimately to the emergence of land rights and native title as political issues in Australia, in part because of the tenacious engagement of Yolngu elders who realised that their culture was threatened by such incursions of the State and that in order to battle them they had to find ways to speak across the cultural abyss between white and Aboriginal people (Williams, 1986). In order to do this they had to understand non-Indigenous epistemologies, law and politics. As a result western-style education has long been considered necessary to Yolngu people in pursuit of their own cultural agendas. However, acceptance of the education has been regulated according to local culture. Like Aboriginal people elsewhere in Australia, the Yolngu want to hear what this education has to say, but they want to decide for themselves how to use it. This includes a determination to maintain local languages and the patterns of life associated with residence in small communities on homelands, while having regular schooling in English, as it is in mainstream schools.

Aboriginal people in the Katherine area have by no means suffered the degree of dislocation and disruption as people in other parts of Australia but they have a different history from the Yolngu. Bulman is a case in point. Here relationships had been built up over generations between cattle ranchers and local groups. While these were certainly not entirely voluntary or favourable to Aboriginal people, they allowed continued contact with country and a compromise way of life that came to be seen as valuable for many. In the late 1960s or 1970s the local cattlemen made it impossible for their Indigenous staff to remain on the property and these people walked off and set up Bulman near to one of their sacred sites. Because of the long history of living alongside other Aboriginal groups but being forced to conduct much daily business with English speakers, the local languages are not much spoken now. Instead, the home language is Kriol, a new Indigenous language with an English lexical base and an Indigenous grammatical structure. The aspirations of this kind of community are commonly more like those of the mainstream. While ownership of their own land and the right to make decisions regarding it will always be important, a good life is seen to include settled employment in jobs that require mainstream education. Although there are intermittent programs attending to the original languages of the Bulman students, there are none that take account of the fact that the home language is Kriol. While this community appears to value and desire mainstream education, their relationship to it is very different from that of the Yolngu.

The Mobile Pre-school program has been running for a relatively short time in its present form; however, it is based on nearly a decade's work by teachers and communities, and its success is intimately tied up with this long lead time and strength of personal commitment and relationships between participants. In the latter half of the 1990s discussion started on the desirability of providing early childhood education to all Aboriginal children, especially those in remote locations. In general, these discussions were initiated by teachers but in all cases proceeded through lengthy and careful negotiations with communities. With collaboration between staff in education and other government departments, and with the active help of community teachers and women's centre staff in some communities, the concept of "pre-school in a box" slowly evolved. It seems that all of the central participants, from the program officer in Darwin (capital city of the Northern Territory) to the mobile pre-school teachers in the regions, as well as some of the community members, have been involved in the program from very early planning days. Before that they all enjoyed positive and longstanding relationships with communities and this depth of history undoubtedly is important to the success of the program.

What appears to have been a fairly informal arrangement at first could be expanded only with substantial funding. This was lobbied for and gained throughout 2000–2001. The difference between this program and previous ones seems to lie in its greater flexibility. One previous model was to equip a vehicle with all the necessary pre-school equipment and a teacher, who then toured remote communities. This meant that each community was seldom visited and had little opportunity to have input to the program or ownership over it. The present scheme provides extensive support to communities who are substantially left to run the daily activities of the pre-school.

In the case of individual communities, there appear to be several ways in which they became participants. In some cases they heard about the program visiting another community and asked about joining in. Sometimes influential people (usually women) in the community instigated discussion of pre-school as a good thing and urged the council or other community body to explore options. In others, the suggestion came from teachers, although communities had ways of electing not to participate. Once a community decided to participate, workshops were held to inform community members about what was involved and seek interested people to act as TSOs. The existing relationships between teaching staff and communities were felt to be crucial in this process as they are in the continuing mentoring relationships between regional pre-school teachers and TSOs.

Significance of the Partnership

Successful government intervention in Aboriginal communities, whether in matters of health, education, social order or employment, is always likely to be fraught. A common criticism is that such services are simply ways in which the State continues to colonise and oppress Aboriginal people by imposing cultural values and behaviours on them that are unwanted and inimical to social and cultural health. For this reason, special care has been taken by those organising this program to take account of the sensibilities of the participating communities. There was no evidence that communities or families were unwilling partners in the program.

This partnership is significant primarily for the success of its articulation of school, home and community sectors in pursuit of better educational outcomes for children. This program demonstrates the truth of arguments in the literature that success depends on sustained mutual collaboration, support and participation of education personnel and families at home and at school (Funkhouser & Gonzales, 1997; Kahne, 1999). There is also a substantial and unusual sharing of decision-making power between teachers, TSOs and the community. Although this program has been in operation for a relatively short time, it depends on exactly those sorts of relationships built up over many years. It demonstrates that such essential relationships cannot be mandated from outside nor built up overnight, but depend on trust and mutual respect which can only be gained over time.

DISTANCE EDUCATION FOR CHILDREN
OF RURAL FAMILIES

Distance education in the Australian context is a well established, formal educational partnership between state-based education departments, schools, parents and students. Through distance education contexts we are able to investigate how teachers and parents (through their dual role of care giver and academic supervisor) establish learning partnerships in situations where decisions about *what* is learnt is dominated by education departments and teachers (that is, *top-down* models) and yet the manner in which learning environments are established is strongly influenced by parents. Distance education fosters learning cultures where the boundaries between "home" and "school" learning become blurred.

Although Distance Education outcomes and syllabus documents are identical to those of regular schools, the context in which learning takes place is quite different from traditional school settings. These home settings tend to range from formal classrooms, where designated areas are created in the home to mirror regular classrooms, to informal arrangements where students learn seamlessly throughout the day through interaction with learning materials and engagement with their supervisor. Learning materials are distributed from a Distance Education Centre (school) each fortnight, with a supervisor (usually a parent) responsible for establishing a learning environment and providing an opportunity for students to complete the designated activities over the two week period. The classroom teacher, who physically could be thousands of kilometres from the student, relies on satellite or radio communications to interact with individuals and small groups of children for approximately two hours per week. Consequently, the home supervisor plays a significant role in the delivery, construction and modification of learning activities. It is important to note that parents are always referred to as "supervisors" rather than "teachers" despite the central role they play in the learning process.

Observations of the Partnership in Operation

This case study is of a Distance Education Centre in a rural setting in the state of New South Wales. We visited the homes of some of the students enrolled with the Centre and attended a residential mini-school associated with it, as well as interviewing teachers at the Centre. It is typical of most distance education services in that teachers at the Centre are provided with standard materials from a central publishing unit from which they then select according to the needs of their students. Materials are sent out to homes where parents supervise the child's learning. In fact the parents are doing much more than making sure the student works through the material. They organise the home and its routines to make learning possible through setting aside a classroom area and fitting learning activities into the child's day in ways that develop good work habits while attending to the needs and personality of the child. They commonly are very active in guiding children through the learning and finding ways to make links with everyday practice and experience which enhance its meaning.

Students are meant to have a weekly telephone conversation with their teacher, but in fact some of the families in this study had no telephone line and no reliable mobile telephone connection. It takes great persistence from everyone involved to maintain the partnership under such circumstances. Even greater difficulties arise when computer or satellite links are used to link partners. Where Internet access is available the download rate is frequently so slow as to make it virtually useless. Audio tapes are the most common form of delayed two-way communication and one family described them as "pure torture." They are used to give feedback, but the delay between doing the work and getting the feedback reduces their efficacy, as well as the fact that such communication is inherently stilted.

Teachers also visit students' homes so as to get a better understanding of their circumstances and how they might affect learning. Because parents are unfamiliar with the appropriate educational goals and standards for children at various ages, particularly with their first or only child, teachers also help by describing what counts as achievement and how fast it should be attained. In this kind of partnership the main roles for the teacher are as assessor of student progress and the nature of their problems and as the planner of learning activities. Teachers appear to be aware of and sensitive about the expectation for them to be expert, even in areas in which they feel less so, such as technology use. They defined a good partnership as one in which the home supervisor could negotiate their role as other than parent while respecting and trusting the teacher and following the teacher's program.

In fact parents do significant work in creating successful learning opportunities for their children. Parents are uniquely well placed to know what kind of routine works best for their child, but there is still a struggle in most cases to make it happen every day in the face of many distractions. The issue of finding learning in naturalistic activities was one that preoccupied many parents and stimulated much innovation. Sometimes these activities related to routine matters in the everyday environment such as working in the kitchen or fencing, but sometimes they related directly to the needs of the students. For instance, we were told of one mother who was innovative in adapting learning objectives and materials to the very specific needs of her autistic son, for whom the standard materials were never suitable.

Many parents say they lack confidence in mathematics and rely very heavily on teacher encouragement in applying the materials provided and developing them for their child. While parents who are novice supervisors try hard to follow the syllabus exactly as it comes from the Centre, experience and the advice of other home supervisors often lead to more innovative approaches. Parents tend to rely on each other to get a sense of whether or not their child is competently moving through the syllabus and achieving the expected outcomes. Impressively, the classroom teachers actually encourage and foster this form of sharing in subtle but quite powerful ways. Organisational strategies identified included the maintenance of a web site and ensuring that home supervisors have the telephone number of other supervisors who are relatively experienced or have children of a similar age.

One of the most effective ways of promoting engagement among supervisors was through centralised mini-schools that brought teachers, parents and children together two or three times per year. These face-to-face sessions are of critical importance to both the supervisor and teacher because they provide opportunities for the stakeholders to engage in conversation about teaching and learning. Mini-schools usually last for about two days and one night, and are held in a central venue such as a school, caravan park, or property of one of the parents. While the students complete educational activities, supervisors meet to discuss their struggles and successes and learn better ways to help their child in the classroom. These mini-school experiences provide an opportunity for students to further their development in all curriculum areas and are an excellent forum for isolated students to socialise. There was a feeling that the mini-schools were an ideal way to meet new parents in an environment outside the school-based activities. Strong friendships developed as a result of the mini-school—with these friendships important both for "friendship's sake" and as a support mechanism for supervisors.

In this case study the material resources mediate the partnership to a very significant degree. They are the physical link between all the stakeholders and the site of contestation between them. These printed materials have inherent difficulties, especially in the remoteness of their content from actual use. Their form and content embody decisions made in a central publishing and curriculum unit remote from all the partners. Teachers told us they would like to have the learning modules in electronic format so that they could modify and recombine elements of them to suit individual student need, but this is not permitted by the central unit. This may be the reason why teachers are quick to encourage supervisors to share ideas and extension activities, while seeming reluctant to take up those ideas themselves.

From the parent perspective, there is confusion and uncertainty over the best use of materials. Apart from the printed module materials, children are supplied with CD-ROMs, some have access to web-based resources, and a large box of physical aids is provided. Parents commented repeatedly that the box of physical learning aids was impressive but that they lacked confidence and guidance in using them. In one case, the box was observed to be buried under piles of paper and old schoolwork. The teachers, on the other hand, assert that the parents do know how to use the materials. This illustrates a problem common to many of the partnerships we studied. Teachers tend to believe that telling parents what to do covers their communication responsibilities and parents are hesitant to reveal what they understand to be their own ignorance by asking for detailed instruction. This is not helped by the use of mathematical terms like "subtraction" without a gloss to illustrate for supervisors what an everyday example might be. While the materials, especially the physical materials, ought to bridge this gap between abstract concepts and experience, manipulables such as blocks and coloured counters have relatively little resonance with the actual environment. The materials are themselves an abstraction caught somewhere between objects that have real world use and symbols on a page.

Numeracy Practices

Although students work with the same mathematics syllabuses as their classmates in regular schools, the home supervisors are able to modify activities to accommodate their own learning contexts, create their own learning materials, or use everyday events and experiences to explain mathematical concepts. Rather than working through the units as given, supervisors may purposely select topics to take advantage of numeracy learning opportunities within the children's home environment. This results, for example, in spatial and measurement activities being integrated into real life contexts long before number concepts are moved beyond pencil-and-paper representations. From a numeracy perspective, mathematical and contextual knowledge and competencies are often developed hand in hand when children learn via distance education. It seems likely also that transforming standard learning materials into contextualised activities may contribute to the development of strategic knowledge as "working mathematically," especially in terms of investigating, conjecturing, using problem solving strategies and applying and verifying (Curriculum Corporation, 1994). An example from the case study, involving a family building a mud brick house, serves to illustrate this point. In the context of making mud bricks, the supervisor and child can *investigate* whether they have enough clay and straw to finish building a wall, *conjecture* as to what would happen if they used more water in the mix, use *problem solving strategies* to make some trial bricks with different proportions of ingredients, and *apply and verify* by calculating the amount of ingredients needed and building the wall.

Context and History of the Partnership

Distance education is a long standing practice in Australian education, enabling children who live in geographically isolated areas to attend school. Nevertheless, there is some evidence to suggest that students in rural and remote settings remain disadvantaged by their location (Ryan, 2001). Dockett, Perry, Howard and Meckley (1999) compared the perceptions of Australian parents in rural and remote locations and those of city parents regarding what is important in young children's transition to school, and found differences associated with the particular effects of geographical isolation, school and class size, the nature of local communities, the form of distance education, and the nature of transition programs. Similarly, du Plessis and Bailey (2000) reported that isolated parents recognize the educational disadvantage their children suffer through their geographical location and that parents thus want realistic and effective resources to support education programs for their children. Provision of electronic materials could go some way towards meeting this need but material infrastructure is in itself an issue for this partnership in the Australian context.

Significance of the Partnership

Some educators have commented that mathematics experiences in and out of school can build on and complement each other (Masingila & de Silva, 2001) when various learning cultures are recognised and celebrated. In distance education the home environment is less of an out-of-school context and can therefore more readily foster and develop students' learning and practice. The home supervisor had an influential role in the construction of the teaching and learning processes being implemented to support young students' numeracy development—with the influence being much more dramatic than the classroom teachers envisaged. The dynamics of the learning environments were significantly different from traditional classroom-based contexts, with the supervisor having the strongest influence over the way in which pedagogical practices and learning outcomes were presented to children. The supervisors were able to establish strong connections between in and out of school engagement and actively attempted to create such contexts even though the blurring of these boundaries created other challenges.

It seemed to be the case that the home supervisors were more committed to developing authentic learning experiences than the students' distance education teachers. Although the teachers had a good understanding of each child's home context there were few examples of curriculum modification to cater for individuals' needs or interests. The dual role of the supervisor (as both parent and teacher) provided opportunities for the reconstruction of learning activities that were embedded in personal contexts and their capacity to access a range of authentic artefacts helped establish powerful and rich learning environments despite their limited knowledge of the curriculum. It could be argued that the distance education teachers did not appreciate the supervisors' willingness to modify learning tasks or recognise their capacity to enhance these learning situations. As Sarason (1995) argued, the structure of schools delineates the nature and scope of parental involvement and can create mismatches between the home and school environments. The failure to recognize parental diversity can cause barriers within these relationships.

DISCUSSION

The aim of this project was to identify the features of effective numeracy education partnerships involving mutual collaboration and participation of children with their families, schools, and communities. We draw on the two case studies presented here, together with data from our national survey of elementary school principals, interviews with education government and non-government education system personnel, and the other case studies we conducted, to discuss these partnership features with respect to the roles and perspectives of stakeholders and ways in which partnerships responded to diversity and disadvantage.

Stakeholder Roles and Perspectives

This issue is concerned with the extent to which partnerships were school-centred, family-centred, or community-centred. Almost all of the 38 programs we identified in our interviews emphasised school-centred perspectives on partnerships, and the most common feature of these programs was the emphasis placed on enhancing *communication* between teachers and parents (cf Epstein, 1995). Distance Education would seem to offer greater than usual prospects for two-way home-school communication, and to some extent this was what we observed. Mini-schools provided an excellent opportunity for parents to share ideas, and home visits helped teachers to understand their students' daily circumstances. However, effective use of communication technologies such as telephone and the Internet—and even very basic technologies such as audio recordings—proved to be problematic in terms of improving learning. Also, although the home supervisors were often very innovative in adapting and creating materials, the teachers rarely incorporated this feedback into their standard practice, thus undermining any potentially positive effects of two-way communication.

We also found in our case studies some evidence of the roles that family and community members may play in supporting children's numeracy development. The Distance Education case study illustrates a distinctly family-centred perspective on educational partnerships. In terms of the analytical framework we introduced earlier in the paper, the parents as supervisors *create supportive learning environments at home* and *recognise home practices that support numeracy development* by finding numeracy learning opportunities in the children's everyday contexts. The distance education teachers also foster this family-centred perspective in a number of ways, for example, by ensuring that supervisors have the telephone number of others parents who are relatively experienced or have children of a similar age. Isolated students and their parents are also brought together through mini-schools and camps to share ideas, and these opportunities for communication between parents in their supervisory roles appear to be crucial in helping them to devise practices that support their children's numeracy learning at home. Distance education seems to offer a unique context for family-school partnerships in that parents take on a role that gives them responsibility for negotiating individual curriculum modifications with their children's teachers and fellow supervisors.

The Mobile Pre-school Pilot Program highlights the significance of community-centred perspectives on educational partnerships, in this case the ways in which community involvement can contribute to *educational reform and curricular enrichment*. The program has had positive outcomes for schools, teachers and communities consistent with the benefits of community-centred education programs identified by Kahne (1999). For example, the local communities benefited because of the new jobs created by the program. This is a direct financial benefit but also a social benefit in that community members are given positions of trust and responsibility, their opinions are listened to (in fact eagerly solicited) and they thus provide a role model and exemplar of one kind of success and one kind of use for education for others in the community.

Response to Diversity and Disadvantage

Around 18% of the programs we identified through our national email survey and interviews with education authorities targeted children from Aboriginal and Torres Strait Islander backgrounds, and some of the programs for which we were able to obtain more detailed information tried to respond to culturally different ways of knowing mathematics. Yet it is important to recognise that cultural difference does not necessarily translate into different numeracy goals and practices: the Mobile Pre-school Pilot Program case study reminds us that Indigenous parents and communities also have legitimate reasons for valuing Western numeracy practices. Significant in this case was the way in which the program was offered to communities on their terms rather than being imposed as a "solution" to a perceived deficiency in children's numeracy learning outcomes. Involving parents and community members as cultural gatekeepers in making decisions about teaching resources and learning activities can also lead to subtle shifts in power relations: the Mobile Pre-school pilot Program was only rolled out to those communities who agreed to have it, thus returning power to the community.

Children in geographically isolated locations were less well served by the numeracy programs (10%) reported in the email survey. Our case study of a Distance Education Centre demonstrated that, in rural contexts, the very nature of the remoteness, isolation and restricted opportunities for communication gives parents (as supervisors) a dominant role in the learning process. In these circumstances supervisors established authentic problem-solving contexts for children to acquire knowledge and skills in situations that were meaningful and relevant to their personal experiences in both school and out-of-school contexts. Distance education, where children are both at school and at home, also has the potential to break down, or at least blur, the traditional barriers that exist between teachers, learners and parents. However, because too many of the partnerships exist out of necessity (remoteness) rather than a desire to create a different set of relations between the participants, the presentation and representation of "mandatory" curriculum content tends to be the dominant discourse of most interactions—with the teachers attempting to ensure that numeracy is adequately covered and understood (by both parents and children alike) while parents (as supervisors) are constantly seeking assurance and attempting to keep up with changes to curricula. It seemed that the distance education teachers were caught in a position of power that prevented them from acknowledging the very substantial contribution of parents to their children's learning.

CONCLUSION

In the field of home, school and community partnerships there is no consistent agreement about the meaning of the terms "partnerships," "parent involvement," and "community involvement." Many different kinds of activities fall within this field. In addition, the stakeholders in these connections between home, school and community may hold conflicting perceptions about numeracy, and about their roles and the roles of other stakeholders. In studying effective partnerships in numeracy education, the importance of relationships, mutual trust, and respect developed over an extended period of time was a theme that emerged from our case study analysis. This essential goodwill cannot be created entirely by funding or targeted programs, and programs such as the Mobile Pre-school Pilot Program and Distance Education in rural Australia owe their success to a long history of cooperation and joint enterprise centred on the welfare and education of children, their families and communities. It was also noteworthy that some of the most effective partnerships we identified for our case studies were not initiated as numeracy programs but took a more holistic approach (cf Hexter, 1990). Our research indicates that building strong home-school-community partnerships around children's learning in general can lay the groundwork for numeracy-specific learning. In culturally diverse communities we would suggest that partnership building is of paramount importance, and should precede—or at least accompany—the introduction of educational programs that seek to initiate children into numeracy practices that are valued but different from those of their home culture.

Finally, we would warn against inferring that the term "partnership" implies that there should be similar contributions from, and roles for, all participants. This was especially salient when considering the roles of parents and teachers in educational partnerships. While we found plenty of evidence that parents genuinely care about their children's education, it was equally clear that not all parents want to be actively involved in all aspects of schooling and many see their role as primarily a supportive one. Perhaps the most productive way forward is to focus on what each participant—parent, teacher, community member—can bring to the partnership that will make best use of their diverse expertise, backgrounds, and interests in supporting the child's numeracy learning.

ACKNOWLEDGEMENT

This research was commissioned by the Australian Government through the Department of Education, Science and Training. The project was funded under the National Strand of the Numeracy Research and Development Initiative. We thank the other members of the research team: Angela Coco, Sandra Frid, Peter Galbraith, Marj Horne, Alex Kostogriz, Trisch Short, Daniel Lincoln, Mohammad Gholam.

REFERENCES

Boaler, J. (1993). The role of contexts in the mathematics classroom: Do they make mathematics more "real"? *For the Learning of Mathematics, 13*(2), 12–17.

Cairney, T. (2000). Beyond the classroom walls: The rediscovery of the family and community as partners in education. *Educational Review, 52*(2), 163–174.

Crozier, G. (2000). *Parents and schools: Partners or protagonists?* Stoke on Trent, UK: Trentham Books.

Curriculum Corporation (1994). *Mathematics—A curriculum profile for Australian schools.* Melbourne: Curriculum Corporation.

Cutler, W. (2000). *Parents and schools: The 150-year struggle for control in American education.* Chicago, IL: University of Chicago Press.

deCarvalho, M. (2001). *Rethinking family-school relations: A critique of parental involvement in schooling.* Mahwah, NJ: Erlbaum.

Department of Education, Training and Youth Affairs (DETYA) (2000). *Numeracy, A priority for all: Challenges for Australian schools.* Canberra: Commonwealth Government of Australia.

Department of Employment, Education, Training and Youth Affairs (DEETYA) (1997). *Numeracy = everyone's business* (Report of the Numeracy Education Strategy Development Conference). Adelaide: Australian Association of Mathematics Teachers.

Dockett, S., Perry, B., Howard, P., & Meckley, A. (1999). What do early childhood educators and parents think is important about children's transition to school? A comparison between data from the city and the bush. In P. Jeffrey & R. Jeffrey (Eds.) *Proceedings of AARE / NZARE Conference.* Retrieved August 31, 2006, from: http://www.aare.edu.au/99pap/per99541.htm.

du Plessis, D., & Bailey, J. (2000). Isolated parents' perceptions of the education of their children. *Education in Rural Australia, 10,* 1–26.

Epstein, J. (1995). School/family/community partnerships: Caring for the children we share. *Phi Delta Kappan, 76*(9), 701–712.

Epstein, J. (2001). *School, family, and community partnerships: Preparing educators and improving schools.* Boulder, CO: Westview Press.

Fine, M. (1993). [Ap]parent involvement: Reflections on parents, power, and urban public schools. *Teachers College Record, 94*(4), 682–729.

Folds, R. (1987). *Whitefella school: Education and Aboriginal resistance.* Sydney: Allen & Unwin.

Funkhouser, J., & Gonzales, M. (1997). *Family involvement in children's education: Successful local approaches.* Washington, DC: US Department of Education.

Goodluck, M. A., Lockard, L., & Yazzie, D. (2000). *Language revitalization in Navajo/ English dual language classrooms.* Retrieved August 31, 2006, from http//:jan. ucc.nau.edu/~jar/LIB/LIB2.html.

Harris, S. (1990). *Two-way Aboriginal schooling: Education and cultural survival.* Canberra: Aboriginal Studies Press for the Australian Institute of Aboriginal and Torres Strait Islander Studies.

Hexter, H. (1990). A description of federal information and outreach programs and selected state, institutional and community models. *Symposium on information resources, services and programs. Background paper number three.* Washington DC: Advisory Committee on Student Financial Assistance.

Horne, M. (1998). Linking parents and school mathematics. In N. Ellerton (Ed.), *Issues in mathematics education: A contemporary perspective* (pp. 115–135). Perth: Mathematics, Science and Technology Education Centre, Edith Cowan University.

James, D., Jurich, S., & Estes, S. (2001). *Raising minority academic achievement: A compendium of education programs and practices. Report.* Washington, DC: American Youth Policy Forum. Retrieved August 31, 2006, from http://www.aypf.org/ publicatons/rmaa/pdfs/Book.pdf

Jordan, C., Ozorco, E., & Averett, A. (2001). *Emerging issues in school, family and community connections. Report.* Austin, TX: Southwest Educational Development Laboratory.

Kahne, J. (1999). Personalized philanthropy: Can it support youth and build civic commitments? *Youth and Society, 30*(3), 367–387.

Katz, Y. (2000). The parent-school partnership: Shared responsibility for the education of children. *Curriculum and Teaching, 15*(2), 95–102.

Keith, N. (1999). Whose community schools? New discourses, old patterns. *Theory into Practice, 38*(4), 225–234.

Malcolm, I. (Ed.) (1999). *Two-way English: Towards more user-friendly education for speakers of Aboriginal English.* East Perth, WA: Education Dept.

Masingila, J., & de Silva, R. (2001). Teaching and learning school mathematics by building on students' out-of-school mathematics practice. In B. Atweh, H. Forgasz, & B. Nebres (Eds.), *Sociocultural research on mathematics education: An international perspective* (pp. 329–344). Mahwah NJ: Erlbaum.

McIntyre, E., Rosebery, E., & Gonzalez, N. (Eds.) (2001). *Classroom diversity: Connecting curriculum to students' lives.* Portsmouth, NH: Heinemann.

Peressini, D. (1998). The portrayal of parents in the school mathematics reform literature: Locating the context for parental involvement. *Journal for Research in Mathematics Education, 29,* 555–572.

Ryan, R. (2001). Human rights, remote Australia, and the VET sector. *Australian Training Review, 40,* 28–29.

Sarason, S. (1995). *Parental involvement and the political principle: Why the existing governance structure of schools should be abolished.* San Francisco, CA: Jossey-Bass.

Saxe, J. (2002). Children's developing mathematics in collective practices: A framework for analysis. *The Journal of Learning Sciences, 11*(2–3), 275–300.

Sfard, A. (2002). The interplay of intimations and implementations: Generating new discourse with new symbolic tools. *The Journal of Learning Sciences, 11*(23), 319–358.

Stanton, R. (1994). Mathematics both ways: A mathematics curriculum for Aboriginal teacher education students. *For the Learning of Mathematics, 14*(3), 15–23.

Wenger, E. (1998). *Communities of practice: Learning, meaning and identity,* Cambridge, MA: Cambridge University Press.

Wenger, E., McDermott, R., & Snyder, W. (2002). *Cultivating communities of practice.* Boston: Harvard Business School Press.

Williams, N. M. (1986). *The Yolngu and their land: A system of land tenure and the fight for its recognition.* Canberra: Australian Institute of Aboriginal Studies.

Willis, S. (1998). Which numeracy? *Unicorn, 24*(2), 32–42.

CHAPTER 3

PEACE, SOCIAL JUSTICE AND ETHNOMATHEMATICS

Ubiratan D'Ambrosio
Pontifícia Universidade Católica de São Paulo, Brazil
and
(Emeritus Professor) State University of Campinas,
São Paulo, Brazil

ABSTRACT

Issues affecting society nowadays, such as national security, personal security, economics, social and environmental disruption, relations among nations, relations among social classes, people's welfare, the preservation of natural and cultural resources, and many others can be synthesized as Peace in its several dimensions: Inner Peace, Social Peace, Environmental Peace and Military Peace. These four dimensions are intimately related. Social Justice, the theme of this book, naturally leads to Social Peace. Although, as I said, the four dimensions of Peace are intimately related, in this chapter I will focus my reflection on Social Justice and how can Ethnomathematics contribute to it.

International Perspectives on Social Justice in Mathematics Education, pages 37–50
Copyright © 2008 by Information Age Publishing
37

THE RESPONSIBILITY OF MATHEMATICIANS
AND MATHEMATICS EDUCATORS

It is widely recognized that all the issues affecting society nowadays are universal, and it is common to blame, not without cause, the technological, industrial, military, economic and political complexes as responsible for the growing crises threatening humanity. Survival with dignity is the *most universal problem facing mankind.*

Mathematics, mathematicians and mathematics educators are deeply involved with all the issues affecting society nowadays. But we learn, through History, that the technological, industrial, military, economic and political complexes have developed thanks to mathematical instruments. And also that mathematics has been relying on these complexes for the material bases for its continuing progress. It is also widely recognized that mathematics is the *most universal mode of thought.*

Are these two universals conflicting or are they complementary? It is sure that mathematicians and math educators, are concerned with the advancement of the most universal mode of thought, that is, mathematics. But it is also sure that, as human beings, they are equally concerned with the most universal problem facing mankind, that is, survival with dignity.

It is absolutely natural to expected that they, mathematicians and math educators, look into the relations between these two universals. That is, mathematicians and math educators look into the most universal problem facing mankind as the most urgent problem to be dealt with. Mathematicians and math educators must accept, as priority, the pursuit of a civilization with dignity for all, in which inequity, arrogance and bigotry have no place. This means, to achieve a world in peace (see Pugwash 1955 and D'Ambrosio 2001). I have no doubt that every mathematician and math educator agree and are concerned with this most universal problem. Their discourse supporting this appeal is, without any doubt, sincere. But once they move into their practice, as mathematicians and math educators, something like a barrier appears and obfuscates their concern. They continue to do what they ever did. For mathematicians, priority is to publish their research in the best journals and for math educators, to propose, theorize and publish methods, which supposedly help teachers to better prepare their students to pass the variety of tests which are imposed on them. And sameness prevails!

Although with a somewhat different focus, discussing individualism in research activity, the late John M. Ziman (2006) described, in a provocative essay, the essence of a familiar attitude:

> To a remarkable degree, the scientist is represented as studying the natural world as if alone in it, served only by mindless assistants who might as

well be replaced by machines. Scientific theories are presented as systems of thought conjured up and tested by that same individual in a further series of single-headed operations. Research results are formulated and treated philosophically as the independent findings of lone explorers, each reporting the evidence of their own eyes and their rational inferences concerning the hidden mechanisms by which these personal percepts might be generated. Our epistemological role models are Robinson Crusoe and Sherlock Holmes, self-sufficient intellectuals to human their human companions Friday and Watson, are mere stooges.

As I said above, Peace must be understood in its multiple dimensions:

- inner peace
- social peace
- environmental peace
- military peace.

My research program is to understand the responsibility of mathematicians and mathematics educators in offering venues for Peace. The Program Ethnomathematics, which will be discussed later in the chapter, is a response to this.

A research program, on mathematics, history, education and on the curriculum, which is an attempt to face the question of responsibility, begins with a reflection on the nature of mathematical behavior. How is mathematics created? How different is mathematical creativity from other forms of creativity? To face these questions there is need of a complete and structured view of the role of mathematics in building up our civilization, hence a look into the history and geography of human behavior.

I emphasize that History not only as a chronological narrative of events, focused in the narrow geographic limits of a few civilizations, which have been successful in a short span of time. The course of the history of mankind can not be separated from the natural history of the planet. The history of civilization has developed in close and increasing interdependence with the natural history of the planet.

About Education, I claim that its major goals are:

- to promote creativity, helping people to fulfill their potentials and raise to the highest of their capability, but being careful not to promote docile citizens. We do not want our students to become citizens who obey and accept rules and codes which violate human dignity.
- to promote citizenship transmitting values and showing rights and responsibilities in society, but being careful not to promote irresponsible creativity. We do not want our student to become bright scientists creating new weaponry and instruments of oppression and inequity.

The big challenge we face is the encounter of the old and the new. The old is present in the societal values, which were established in the past and are essential in the concept of citizenship. And the new is intrinsic to the promotion of creativity, which points to the future. The strategy of education systems to pursue these goals is the curriculum. Curriculum is usually organized in three strands: objectives, contents, and methods. This Cartesian organization implies accepting the social aims of education systems, then identifying contents that may help to reach the goals and developing methods to transmit those contents.

THE POLITICAL DIMENSIONS
OF MATHEMATICS EDUCATION

The discussion on the objectives of Mathematics Education or, in other words, on "Why teach mathematics?" is regarded as the political dimension of education, but very rarely we see mathematics content and methodology been examined with respect to this dimension (see Sriraman & Törner, 2007). Indeed, some educators and mathematicians claim that content and methods in mathematics have nothing to do with the political dimension of education. Even more disturbing is the possibility of offering our children a world convulsed by wars. Because mathematics conveys the imprint of Western thought, it is naïve not to look into a possible role of mathematics in framing a state of mind that tolerates war. As argued above, our major responsibility, as mathematicians and mathematics educators, is to offer venues of peace (D'Ambrosio 1998).

There is an expectation about our role, as mathematicians and mathematics educators, in the pursuit of peace. Anthony Judge (2000), when director of communications and research of the Union of International Associations, expressed how we, mathematicians, are seen by others:

> Mathematicians, having lent the full support of their discipline to the weapons industry supplying the missile delivery systems, would claim that their subtlest thinking is way beyond the comprehension of those seated around a negotiating table. They have however failed to tackle the challenge of the packing and unpacking of complexity to render it comprehensible without loss of relationships vital to more complex patterns. As with the protagonists in any conflict, they would deny all responsibility for such failures and the manner in which these have reinforced unsustainably simplistic solutions leading to further massacres.

I see my role as an educator and my discipline, mathematics, as complementary instruments to fulfill commitments to mankind. To make good use of these instruments, I must master them, but I also need to have a critical view of their potentialities and of the risk involved in misusing them. This is my professional commitment.

It is difficult to deny that mathematics provides an important instrument for social analyses. Western civilization entirely relies on data control and management. "The world of the twenty-first century is a world awash in numbers" (Steen 2001, 1). Social critics will find it difficult to argue without an understanding of basic quantitative mathematics.

Since the emergence of modern science, enormous emphasis has been placed on the rational dimension of man. Recently, multiple intelligences, emotional intelligence, spiritual intelligence, and numerous approaches to cognition, including new developments in artificial intelligence, challenge this. In mathematics education, this challenge is seen in the exclusive emphasis given to skill and drilling, as defended in some circles of mathematicians and mathematics educators.

I argue against the excessive emphasis on the quantitative, which may be detrimental to the equally important emphasis on the qualitative. My proposal of *literacy, matheracy,* and *technoracy,* discussed below, is an answer to my criticism of the lack of equilibrium. *Literacy* is a communicative instrument and, as such, includes what has been called quantitative literacy or numeracy. This is very much in line with the mathematics learned from the Egyptians and Babylonians, but not central in Greco-Roman civilization nor in the High Middle Ages. It was incorporated into European thought in the Lower Middle Ages and it was essential for mercantilism and for the development of modern science. Indeed, it became the imprint of the modern world. In contrast, *matheracy* is an analytical instrument, as proposed by classical Greek mathematicians (for example, in Plato's *Republic*). I will return to this subsequently.

It is an undeniable right of every human being to share in all the cultural and natural goods needed for material survival and intellectual enhancement. This is the essence of the United Nations' *Universal Declaration of Human Rights* (UN 1948) to which every nation is committed. The educational strand of this important profession on the rights of mankind is the *World Declaration on Education for All* (UNESCO 1990) to which 155 countries are committed. Of course, there are many difficulties in implementing United Nations resolutions and mechanisms. But as yet this is the best instrument available that may lead to a planetary civilization, with peace and dignity for all mankind. Regrettably, mathematics educators are generally unfamiliar with these documents.

THE ETHICAL DIMENSION OF MATHEMATICS EDUCATION

It is not possible to relinquish our duty to cooperate, with respect and solidarity, with all the human beings who have the same rights for the preservation of good. The essence of the ethics of diversity is respect for, solidarity with, and cooperation with the other (the different). This leads to quality of life and dignity for all.

It is impossible to accept the exclusion of large sectors of the population of the world, both in developed and undeveloped nations. An explanation for this perverse concept of civilization asks for a deep reflection on colonialism. This is not to place blame on one or another, not an attempt to redo the past. Rather, to understand the past is a first step to move into the future. To accept inequity, arrogance, and bigotry is irrational and may lead to disaster. Mathematics has everything to do with this state of the world. A new world order is urgently needed. Our hopes for the future depend on learning - critically - the lessons of the past.

We have to look into history and epistemology with a broader view. The denial and exclusion of the cultures of the periphery, so common in the colonial process, still prevails in modern society. The denial of knowledge that affects populations is of the same nature as the denial of knowledge to individuals, particularly children. To propose directions to counteract ingrained practices is the major challenge of educators, particularly mathematics educators. Large sectors of the population do not have access to full citizenship. Some do not have access to the basic needs for survival. This is the situation in most of the world and occurs even in the most developed and richest nations.

Let me discuss the proposal of a new concept of curriculum, synthesized in three strands: *literacy, matheracy,* and *technoracy* (D'Ambrosio 1999b). The three provide, in a critical way, the communicative, analytical and technological instruments necessary for life in the twenty-first century. Let me discuss each one.

Literacy is the capability of processing information, such as the use of written and spoken language, of signs and gestures, of codes and numbers. Clearly, reading has a new meaning today. We have to read a movie or a TV program. It is common to listen to a concert with a new reading of Chopin. Also, socially, the concept of literacy has gone through many changes. Nowadays, reading includes also the competency of numeracy, the interpretation of graphs and tables, and other ways of informing the individual. Reading even includes understanding the condensed language of codes. These competencies have much more to do with screens and buttons than with pencil and paper. There is no way to reverse this trend, just as there has been no successful censorship to prevent people from having access to books in the past 500 years. Getting information through the new media supersedes the use of pencil and paper and numeracy is achieved with calculators. But, if dealing with numbers is part of modern literacy, where has mathematics gone?

Matheracy is the capability of inferring, proposing hypotheses, and drawing conclusions from data. It is a first step toward an intellectual posture, which is almost completely absent in our school systems. Regrettably, even conceding that problem solving, modeling, and projects can be seen in some mathematics classrooms, the main importance is usually given to numeracy, or the manipulation of numbers and operations. Matheracy is closer to the way mathematics was present both in classical Greece and in indigenous cultures. The concern was not with counting and measuring but with divination and philosophy. Matheracy, this deeper reflection about man and society, should not be restricted to the elite, as it has been in the past.

Technoracy is the critical familiarity with technology. Of course, the operative aspects of it are, in most cases, inaccessible to the lay individual. But the basic ideas behind technological devices, their possibilities and dangers, the morality supporting the use of technology, are essential issues to be raised among children at a very early age. History show us that ethics and values are intimately related to technological progress.

The three together constitute what is essential for citizenship in a world moving swiftly toward a planetary civilization.

THE PROGRAM ETHNOMATHEMATICS

A response to the responsibility of mathematicians and mathematics educators and a realization of this new concept of curriculum is the Program Ethnomathematics. To build a civilization that rejects inequity, arrogance, and bigotry, education must give special attention to the redemption of peoples that have been, for a long time, subordinated and must give priority to the empowerment of the excluded sectors of societies.

The *Program Ethnomathematics* contributes to restoring cultural dignity and offers the intellectual tools for the exercise of citizenship. It enhances creativity, reinforces cultural self-respect, and offers a broad view of mankind. In everyday life, it is a system of knowledge that offers the possibility of a more favorable and harmonious relation between humans and between humans and nature (D'Ambrosio 1999a).

The Program Ethnomathematics offers the possibility of harmonious relations in human behavior and between humans and nature. it has; intrinsic to it; the *ethics of diversity*:

- respect for the other (the different);
- solidarity with the other;
- cooperation with the other.

Let me elaborate on the genesis of this research program, which has obvious pedagogical implications.

An important question, frequently asked: is Ethnomathematics research or practice?

I see Ethnomathematics arising from research, and this is the reason for calling it the Program Ethnomathematics. But equally important, indeed what justifies this research, are the implications for curriculum innovation and development, teaching, teacher education, policy making and the effort to erase arrogance, inequity and bigotry in society.

For almost three decades, I have been formally involved with Pugwash Movement and the pursuit of peace (in all four dimensions: individual, social, environmental and military) (Pugwash 1955). A lecture of the History of Mankind makes it clear that Mathematics is central in all these dimensions. There is no need to elaborate on this.

An insight is gained by looking into non-Western civilizations. I base my research on established forms of knowledge (communications, languages, religions, arts, techniques, sciences, mathematics) and in a theory of knowledge and behavior which I call the "cycle of knowledge." This theoretical approach recognizes the cultural dynamics of the encounters, based on what I call the "basin metaphor." All this links to the historical and epistemological dimensions of the Program Ethnomathematics, which can bring new light into our understanding of how mathematical ideas are generated and how they evolved through the history of mankind. It is fundamental to recognize the contributions of other cultures and the importance of the dynamics of cultural encounters.

Culture, understood in its widest form, which includes art, history, languages, literature, medicine, music, philosophy, religion, science, technology, is characterized by shared knowledge systems, by compatible behavior and by acceptance of an assemblage of values. Research in ethnomathematics is, necessarily, transcultural and transdisciplinarian. The encounters of cultures are examined in its widest form, to permit exploration of more indirect interactions and influences, and to permit examination of subjects on a comparative basis.

Although academic mathematics developed in the Mediterranean Basin, expanded to Northern Europe and later to other parts of the World, it is difficult to deny that the codes and techniques which were developed, such as measuring, quantifying, inferring and the emergence of abstract thinking, as strategies to express and communicate the reflections on space, time, classifying, comparing, which are proper to the human species, are contextual. Clearly, in other regions of the World, other context give origin to different codes and techniques developed as strategies to express and communicate the reflections of a different spatial context, a different time perception, and different ways of classifying and comparing. These are, obviously, contextual.

At this moment, it is important to clarify that my view of ethnomathematics should not be confused with ethnic-mathematics, as it is mistakenly understood by many. This is the reason why I insist in using Program Ethnomathematics, which tries to understand and explain the various system of knowledge, such as mathematics, religion, culinary, dressing, football and several other practical and abstract manifestations of the human species in different contextual realities. Of course, the Program Ethnomathematics was initially inspired by recognizing ideas and ways of doing that reminds us of Western mathematics. What we call mathematics in the academia is a Western construct. Although dealing with space, time, classifying, comparing, which are is proper to the human species, the codes and techniques to express and communicate the reflections on these behaviors is undeniably contextual. I got an insight into this general approach while visiting other cultural environments, during my work in Africa, in practically all the countries of continental America and the Caribbean, and in some European environments. Later, I tried to understand the situation in Asia and Oceania, although with no field work. Thus, came my approach to Cultural Anthropology (curiously, my first book on Ethnomathematics was placed by the publishers in a collection of Anthropology).

To express these ideas, which I call a research program, I created a neologism, *ethno* + *mathema* + *tics*. This caused much criticism, because it does not reflect the etymology of "mathematics." Indeed, "mathematics" is not composed, it is a neologism, with Greek origin, introduced in the XIV century. It is not mathema+tics. The idea of organizing these reflections occurred to me while attending International Congress of Mathematicians ICM 78, in Helsinki. In playing with Finnish dictionaries (to play with dictionaries is a favorite pastime), I was tempted to write *alustapasivistykselitys* for the research program. Bizarre! So, I believed Ethnomathematics would be more palatable.

I understand that there are immediate questions facing World societies and education, particularly mathematics education. As a mathematics educator, I address these questions. Thus, the Program Ethnomathematics links to the study of curriculum, and to my proposal for a modern *trivium*: literacy, matheracy and technoracy.

The pursuit of Peace, in all four dimensions mentioned above, is an urgent need. Thus, the relation of the Program Ethnomathematics with Peace, Ethics and Citizenship. These lines of work in mathematics education link, naturally, to the pedagogical and social dimensions of the Program Ethnomathematics.

As I said above, it is important to insist that the Program Ethnomathematics is not ethnic mathematics, as some commentators interpret it. Of course, one has to work with different cultural environments and, as an ethnographer, try to describe mathematical ideas and practices of other cultures. This is a style of doing ethnomathematics, which is absolutely necessary. These cultural environments include not only indigenous populations, but labour and artisan groups, communities in urban environment and in the periphery, farms, professional groups. These groups develop their own practices, have specific jargons and theorize on their ideas. This is an important element for the development of the Program Ethnomathematics, as important as the cycle of knowledge and the recognition of the cultural encounters.

Basically, investigation in ethnomathematics start with three basic questions:

1. How are *ad hoc* practices and solution of problems developed into methods?
2. How are methods developed into theories?
3. How are theories developed into scientific invention?

It is important to recognize the special role of technology in the human species and the implications of this for science and mathematics. Thus, the need of History of Science and Technology (and, of course, of Mathematics) to understand the role of technology as a consequence of science, but also as an essential element for furthering scientific ideas and theories. (D'Ambrosio 2004).

Once recognized the role of technology in the development of mathematics, reflections about the future of mathematics propose important questions about the role of technology in mathematics education. Besides these more immediate concerns, there are long term concerns. Of course, they are related. Hence, the importance of linking with Future Studies. And, in particular, with Distance Education.

Reflections about the presence of technology in modern civilization leads, naturally, to question about the future of our species. Thus, the importance of the emergent fields of Primatology and Artificial Intelligence, which lead to a reflection about the future of the human species. Cybernetics and human consciousness lead, naturally, to reflections about fyborgs (a kind of "new" species, *i.e.*, humans with enormous inbuilt technological dependence). Our children will be fyborgs when, around 2025, they become decision makers and take charge of all societal affairs. Educating these future fyborgs calls, necessarily, for much broader concepts of learning and teaching. The role of mathematics in the future is undeniable. But what kind of mathematics?

To understand how, historically, societies absorb innovation, is greatly aided by looking into fiction literature (from iconography to written fiction, music, and cinema). It is important to understand the way material and intellectual innovation permeates the thinking and the myths, and the ways of knowing and doing of non-initiated people. In a sense, how new ideas vulgarise, understanding vulgarise as making abstruse theories and artefacts easier to understand in a popular way.

How communities deal with space and time, mainly to understand the the sacralization of chronology and topology in history, is also central.

We have to look into the cultural dynamics of the encounter of generations (parents and teachers and youth). This encounter is dominated by mistrust and cooptation, reflected in testing and evaluation practices, which dominate our civilization. In mathematics education, this is particularly disastrous. Mathematics is, usually, seen by youth as uninteresting, obsolete and useless way. And they are right. Much of is in the traditional curriculum is uninteresting, obsolete and useless.

Resources to testing is the main argument to justify current math contents. The claims of the importance of current math contents are fragile. Myths surround these claims.

It is important to understand children and youth behavior and their expectations. History gives us hints on how periods of great changes affect the relations between generations. Most interesting is the analysis of youth movements after WWII and Viet Nam War. Particularly 1968.

Regrettably, education, in general, is dominated by a kind of "corporate" attitude, in the sense that there is more concern with the subjects taught than with the children. This is particularly true with Mathematics Education. There is more concern with attaining pre-decided goals of proficiency, which favours sameness and may lead to the promotion of docile citizens and irresponsible creativity. Tests are the best instruments to support this corporate aspect of education. Tests penalize creative and critical education, which leads to intimidation of the new and to the reproduction of this model of society.

THE CRITICS OF ETHNOMATHEMATICS

What to say to critics who dismiss ethnomathematics as political correctness gone too far?

It is difficult to deny that mathematics, as well as education in general, are the arms of a political and ideological posture. Ethnomathematics is no different. Yes, ethnomathematics is political correctness. If proposing a pedagogical practice which aims at eliminating truculence, arrogance, intolerance, discrimination, inequity, bigotry and hatred, is labeled as going too far, what to say?

Is it questionable to refer to truculence, arrogance, intolerance, discrimination, inequity, bigotry, hatred, when discussing mathematics? Then the question is not pedagogical, but historical. As I said above, the historiography of mathematics has been very conservative and biased. Of course, both pedagogical and historical issues are related (see D'Ambrosio, 1998b).

What to say to critics who charge that ethnomathematics does little to advance students' knowledge and understanding of mathematics?

It is clear that traditional teaching of mathematics is not satisfying. Testing and assessment is part of the traditional teaching. The alarming results of tests are the result of a very poor education, which is performed in the traditional methods and curricula. Ethnomathematics do not reach a substantial student population and do not have any effect in the bad results of testing. Measures to "tighten" traditional teaching, hoping to get better results in tests and assessment, are nothing less than disastrous. Countries which are model of traditional teaching and are proud of their systems, are the most vulnerable.

It is clear that reinforcing sameness is not the answer to vulnerability. Although history tells this, and as examples I mention the fall of the Roman empire, the collapse of the Third Reich, the fiasco of the Soviet interference in Afghanistan, the demolition of the Berlin Wall, and others. Sameness, like fundamentalism, lacks creativity to counter vulnerability. Tightening measures lead to worsening effects.

This is particularly true in education. Insisting in obsolete, uninteresting and useless mathematics education, will not avoid its rejection by the new generations. On the other hand, by focusing on individual dignity, recognizing the previous knowledge of the individual and of her/his culture [ethnomathematics], we can prepare the most fertile ground for building up new knowledge [mathematics].

It is an important step in education to recognize that all forms of knowledge, both ethnomathematics and mathematics as well, have limitations. So, it is natural to look for new communicative and analytic instruments. This is why history of mathematics and ethnomathematics should be together. Every advance in mathematics is related to overcoming difficulties in doing or explaining something. The advancement of knowledge and understanding of mathematics, once the ground is fertile, is a matter of motivation. Has much more to do with the overall goals and objectives of mathematics education. Why to deny ethnomathematics, which is clearly alive in the professions, in communities, in extant cultures and in cultural history? Who is afraid of it?

When we teach ethnomathematics of other cultures, for example, the mathematics of ancient Egypt, the mathematics of the Mayas, the mathematics of basket weavers of Mozambique, the mathematics of Jequitinhona ceramists, in Minas Gerais, Brazil, and so and so, *it is not because it is impor-*

tant for children to learn any of these ethnomathematics. It is because there is a deep educational reason for this.

Like in language, if we domain only one language, we are less equipped to succeed in the modern World if we have some proficiency in other languages. And it is a known fact that knowing other languages is a positive factor in bettering the domain of one's own language.

The main reasons for ethnomathematics in the curriculum are:

1. to de-mystify a form of knowledge [mathematics] as being final, permanent, absolute, unique. There is a current misperception in societies, very damaging, that those who perform well in mathematics are more intelligent, indeed "superior" to others. This erroneous impression given by traditional mathematics teaching is easily extrapolated to religious, ideological, political, racial creeds;
2. to illustrate intellectual achievement of various civilizations, cultures, peoples, professions, gender. Mathematics is absolutely integrated with Western civilization, which conquered and dominated the entire world. The acceptance, forced or voluntary, of Western knowledge, behavior and values, can not be associated with ideas like "the winner is the best, the losers are to be discarded." More than any other form of knowledge, mathematics is identified with winners. This is true in history, in the professions, in everyday life, in families, in schools. The only possibility of building up a planetary civilization depends on restoring the dignity of the losers and, together, winners and losers, moving into the new. This requires respect for each other. Otherwise, the efforts will be from the losers to become winners, and from the winners to protect themselves from the losers, thus generating defensive confrontation.

Ethnomathematics practices in school favour respect for the other and solidarity and cooperation with the other. It is thus associated with the pursuit of PEACE. The main goal of Ethnomathematics is bulding up a civilization free of truculence, arrogance, intolerance, discrimination, inequity, bigotry and hatred.

REFERENCES

D'Ambrosio, Ubiratan. 1998. "Mathematics and Peace: Our Responsibilities." *Zentralblatt für Didaktik der Mathematik/ZDM*, 30(3): 67–73.

D'Ambrosio, Ubiratan and Marianne Marmé, 1998b: Mathematics, peace and ethics. An introduction, *Zentralblatt für Didaktik der Mathematik/ZDM*, Jahrgang 30, Juni 1998, Heft 3.

D'Ambrosio, Ubiratan. 1999a. "Ethnomathematics and its First International Congress." *Zentralblatt für Didaktik der Mathematik, ZDM.* 31(2): 50–53.

D'Ambrosio, Ubiratan. 1999b. "Literacy, Matheracy, and Technoracy: A Trivium for Today." *Mathematical Thinking and Learning,*1(2): 131–53.

D'Ambrosio, Ubiratan. 2001. "Mathematics and Peace: A Reflection on the Basis of Western Civilization." *Leonardo,* 34(4): 327–332.

D'Ambrosio, Ubiratan 2004. "Ethnomathematics and its Place in the History and Pedagogy of Mathematics," *Classics in Mathematics Education Research,* eds. Thomas P. Carpenter, John A. Dossey and Julie L. Koehler, National Council of Teachers of Mathematics, Reston, VA, 2004; pp. 194–199.

Judge, Anthony. 2000. "And When the Bombing Stops: Territorial Conflict as a Challenge to Mathematicians." *Union of International Associations.* Retrieved January 25, 2002, at http://www.uia.org/uiadocs/mathbom.htm

Pugwash 1955. *Pugwash Conferences on Science and World Affairs.* Retrieved January 25, 2002, at http://www.pugwash.org/

Sriraman, Bharath & Törner, Günter. 2007. Political Union/Mathematics Education Disunion: Building Bridges in European Didactic Traditions. (in press) In L. English (Ed.). *Handbook of International Research in Mathematics Education (2nd ed.).* Mahwah, NJ: Erlbaum.

Steen, Lynn Arthur, ed. 2001. *Mathematics and Democracy: The Case for Quantitative Literacy.* Princeton, NJ: National Council on Education and the Disciplines.

United Nations. 1948. *Universal Declaration of Human Rights.* Retrieved January 25, 2002, at http://www.un.org/Overview/rights.html

UNESCO. 1990. *World Declaration on Education for All.* Retrieved January 25, 2002, at http://www.unesco.org/education/efa/ed_for_all/background/jomtien_declaration.shtml

Ziman, John. 2006. No Man Is An Island, *Journal of Consciousness Studies,* vol. 13, nº 5, 2006; pp. 17–42; p. 17.

CHAPTER 4

MATHEMATICAL MARGINALISATION AND MERITOCRACY

Inequity in an English Classroom

Andrew Noyes
University of Nottingham, UK

ABSTRACT

In this chapter I explore the structuring of English children into learning and life trajectories and the part that mathematics has in this process. Using case reports of two ten-year olds in their final year of primary school education, I examine how broader family social milieu impact upon mathematics learning trajectories. Stacey and Edward live not far from one another in a city in the midlands of England and have been in the same class from age 5 to 11 yet their social distance is considerable. Through the mobilisation of various classed and classifying responses to school mathematics they have developed two very different perspectives on the value of mathematical study. This examination of mathematical marginalisation and misrecognised meritocracy raises questions about the extent to which teachers can disrupt such processes.

International Perspectives on Social Justice in Mathematics Education, pages 51–68
Copyright © 2008 by Information Age Publishing
51

INTRODUCTION

Children from marginalised groups in our society are learning mathematics in most English comprehensive schools. However, in spite of our apparently inclusive school system—currently under threat by the neo-liberal shift to establishing school markets—classifying work happens in school on a daily basis. England has a strong class history and although the traditional distinctions of the "working" and "middle" classes are clumsy and outdated social hierarchies continue to impact upon learners of mathematics in schools. Bourdieu (1998, 86), writing from a French perspective, argued that social classes do not objectively exist but that the process of classification is ongoing, "something to be done," and it is aspects of these processes that I want to explore here. I am particularly interested in the ways in which school mathematics contributes to this social stratification. We know that mathematics tends to act as a social filter (Howson and Wilson 1986; Davis 1993), particularly through the examinations that ascribe *final judgements* in learners:

> Often with a psychological brutality that nothing can attenuate, the school institution lays down its final judgements and its verdicts, from which there is no appeal, ranking all students in a unique hierarchy of all forms of excellence, nowadays dominated by a single discipline, mathematics. (Bourdieu 1998, 28)

However, the arrival at this point is the culmination of many years of gradual distinction; of subtle effects on the learning trajectories of learners which culminate is significant difference, much of which is related to social background and the economic, cultural and social resources of children and their families. Much of this is as a result of structural limitations in the English education system. For instance, we have had for many years under recruitment of mathematics teachers and these have then been dispersed unevenly between types of schools (Noyes 2006). The impact that this has is cumulative upon schools in more challenging circumstances as they cannot recruit enough teachers whilst those in more affluent suburbs can pick the best of the new recruits. Moreover, the National Curriculum, designed to achieve the goals of the "old humanists" and "industrial trainers" (Ernest 1992) has done little for those groups less likely to achieve the magical grade C at the end of their compulsory mathematics education.

Heymann, writing of the German mathematics education could equally well be describing the UK when he says that "almost everything that goes beyond the standard subject matter of the first seven years of schooling can be forgotten without the persons involved suffering from any noticeable disadvantages" (Heymann 2003, 86). Whilst this may be true in one sense - regarding curriculum content—it is certainly not the case that lack of success in secondary mathematics does not lead to disadvantage. Mathematics has formatting power (Skovsmose 1998), not only through its application in science, business and more generally in society but through the way it is used to organise people. This organisation might be the explicit filtering work of school mathematics (e.g. through access to further educational and work opportunities) or hidden in the daily mathematics practices of life (e.g. through understanding of finance and credit arrangements and the implications that such things have upon economic well-being).

My interest in this paper is not about different schools with supposed homogenous groups of advantaged or disadvantaged students because my concern is with how mixed groups of learners get sifted through their mathematics education. It is clear that differences between schools in quite different social milieu might have an effect upon mathematics learning. However, what happens in a particular school for children with quite different social, economic and cultural resources? How does mathematics education function for them? In secondary schools (and increasingly now at the primary level) one way in which this happens is through the setting of children into ability groups from aged 11. In the UK this is predicated upon performance in National Tests and we know from the work of Cooper and Dunne (Cooper 1998; Cooper and Dunne 2000) that this will tend to disadvantage those with lower levels of cultural and linguistic capital (Zevenbergen 2001). One local teacher recently noted in her Masters research report that all of her bottom set were on the Free School Meals (FSM) register (taken as the default measure of socio-economic status) whereas there were very small numbers in the top sets in similar circumstances. If setting by FSM would yield similar group composition to setting by "ability" we must reconsider the social construction of ability (Gillborn and Youdell 2001) and the potential inequity of ability grouping (Boaler, Wiliam et al. 1998; Zevenbergen 2005). This debate continues in the UK with both the main political parties using setting as part of the "standards" agenda rhetoric and a potential vote-winner for "middle England." There is also evidence to suggest that the different supply of teacher quality across the education system is also replicated within schools: Nardi and Steward's (2003) study of T.I.R.E.D. mathematics highlighted the perception amongst some learners that better groups (i.e. whose with fewer FSM pupils?) were given teachers who were perceived to be better. There is a type of fractal similarity of these processes between schools, between groups and then within groups themselves, with some children benefiting more from the teachers instruction than others.

It is this lower level of analysis, the classroom, that I want to consider here by looking into a primary school classroom in which children are not grouped for mathematics. At the time this data was collected many schools were imitating the secondary school penchant for ability grouping but this school was retaining an apparently all-ability approach with children working in mixed groups. These two children had worked together in the same class for 6 years; they have had the same teachers, the same peers and so on. They both live in single parent families but that is where the similarity ends for the capital resources (economic, cultural and social) were quite different and here again Bourdieu helpfully explains how, in the light of this difference, an apparently fair, all-ability teaching approach might not be so equitable:

> ... to penalize the underprivileged and favour the most privileged, the school has only to neglect, in its teaching methods and techniques and its criteria when making academic judgements, to take into account the cultural inequalities between the different social classes. In other words, by treating all pupils, however unequal they may be in reality, as equal in rights and duties, the educational system is led to give its *de facto* sanction to initial cultural inequalities. The formal equality which governs pedagogical practice is in fact a cloak for a justification of indifference to the real inequalities with regard to the body of knowledge taught or rather demanded. (Bourdieu 1974, 37)

This is a really important idea that has considerable implications for mathematics educators; assuming equality is to reinforce inequality when teaching. I will return to this in the closing discussion.

The data reported here aim to ground this idea in the mathematical work of two children: Edward and Stacey. It comes from longitudinal case studies that included school participant-observation, interviews with teachers, children and their parents and video diaries (Noyes 2004) compiled by the children during their final year in primary school. Although they are simply two case studies they are part of a larger group for whom similar processes were at work and using the fractal similarity metaphor I would contend that such case do in fact tell us a lot about the "big picture." What is interesting for these two children is the role of the parental aspirations and how their investment of inherited cultural and social resources in the schooling process accrue different gains for Edward and Stacey. To make sense of these two learners' mathematical trajectories it is insufficient to look only at their classroom behaviours so the data reported here are as much about their family context as they are about mathematics.

Stacey

Stacey's mathematics attainment was amongst the lowest in the group. Despite her academic record she proved to be very capable of reflecting upon her personal and social circumstances, being able to articulate her thoughts and feelings clearly, although sometimes with muddled speech. Her "style of speech and accent," being unsophisticated in comparison to the other children, and regularly containing spoken errors, suggested that she came from a lower income family (Bourdieu 1974, 39). Her relatively low level of linguistic capital put her at a disadvantage in the schooling field because although schools require linguistic "ability," which is not really ability but the result of family socialisation, they do not give it (Bourdieu 1973, 80). So only those children who come already endowed with such capital are in a position to make the most of the opportunities schools purport to "offer" equitably to all children (Bourdieu 1974).

Stacey lived at home with her mum and older sister, not far from the school in a small semi-detached property. She often spoke of her relationship to her older sister, usually in negative terms, describing what Laura had said or done to her. The impact on her mathematical development and developing attitudes to school and learning can be seen in a number of places. In her first video diary entry Stacey made it clear what she thought about, or what she wanted me to think she thought about, mathematics:

Stacey's diary: Maths...I hate it! I hate it! I hate it! Three things about maths...boring, boring, boring!

[Follow-up interview]

Andy: Would you like to explain a bit about that?

Stacey: I don't like maths cause it's hard

Andy: Could you be more specific?

Stacey: Well I can't really time [multiply] that well...my sister calls me a dumb-ass, excuse me.

Andy: Is your sister good at maths?

Stacey: I don't know, she doesn't go to school ever.

Andy: How old is she?

Stacey: Sixteen

Andy: What year is she in?

Stacey: Er...year 10 or 11

Andy: So she just doesn't go to school?

Stacey: Well she does sometimes...it depends whether my mum makes her.

Andy: What does she do when she doesn't go to school?

Stacey: She hangs around with her friends.

Andy: Who are they?

Stacey: Tammy Watson, but they ran away twice so they are not allowed to make contact with each other so Laura's getting mardy and she pretends to be ill.

The move to talking about her sister effectively shifted my attention away from mathematics through her admission that her sister calls her a "dumb-ass." Unlike Edward, Stacey was not comfortable talking about her mathematics learning, partly due to what mathematics made her feel like. Here again the influence of the older sister is apparent:

Diary entry: The thing that makes maths hard for me is that I don't think I'm really good at it...erm...I have to say this prufully, I mean trufully...erm...I know what everyone's thinking, that...I'm the dumbest kid in the class...and...me and Sonya really need desperate help. I'm not saying that she's bad or anything but me and Sonya need really desperate help.

[Follow-up interview]

Andy: Do you really think that everybody thinks you're the dumbest kid in the class? [Stacey nods] How do you know that?

Stacey: Because every time I put my hand up I normally like, roughly get it wrong.

Andy: But you do get some things right don't you? Do you still think that Mrs Clarkson doesn't like you?

Stacey: She didn't like my sister as well. She kept on shouting at my sister and everything.

Andy: She doesn't shout at you though does she?

Stacey: Well not roughly. She kind of like shouts a bit then after I've done my spelling test she goes [teacher voice] "I just want the best for you."

Andy: Do you think that she does want the best for you?

Stacey: No.

Andy: What do you think that she does want then?

Stacey: Erm to be like Matt cause everyone thinks that like Matt's the teachers pet...my sister says that I might have to go to a school that helps people that like need help...but I'm scared of that...don't know why I'm telling you that.

Andy: I would think that you are scared of that. Did she say where you would go?

Stacey: No she said like this school where all the dumb kids have to go and you never see your mum or dad again.

Stacey did often "roughly, get it wrong" in lessons that I observed, and although she didn't appear to mind much in the classroom, it was a cause of anxiety, confirming her self-designation as "the dumbest kid in the class."

Here Stacey is subject to her own and others' classifying work and there is a certain amount of emotional violence to this process. She evidently felt that she would not be able to please the teacher if she could not become like Matt, and her mocking use of the teacher's voice adds further weight to her feeling that her best was not good enough. Stacey's simple understanding of how teacher rhetoric can disguise subtle forms of favouritism reveals her sense of how the school system is unjust. She knows that the school field values highly successes like Matt and she also knows that despite her best efforts she cannot be like him. Carr and Kurtz-Costes (1994, 264) point out that "teachers' subtle messages to children about their abilities influence not only their view of themselves, but also their classmates' expectations for their academic performance." Even in the normal interactions of the year six classroom such "subtle messages" give children a clear idea of their relative position in the group. So Stacey reveals how the teacher uses language to strengthen the children's intellectual, and therefore cultural and social positions, within the group.

Stacey also articulated how the two sisters have to compete for their mum's time. This was due in part to the long shifts that mum worked in her job, sometimes not getting home until eight in the evening. The kinds of work that Stacey's mum could get with her educational qualifications were limited, highlighting another process whereby Stacey's social context structured her life. Stacey would often speak of the cool response that she got from her exhausted mum and these economic conditions resulting from low educational achievement now lead to reduced support for Stacey's schooling. Stacey's mum, by her own account, "did crap at school":

> **Stacey's mum:** My school? I went to quite a few schools 'cause my parents separated. I went to primary school...I went to two, one was a church school and one wasn't...failed my 11+...I've been to a fair few schools and...I never went to school at all in my last year.
>
> **Andy:** Really, and did you get away with that?
>
> **Mum:** Yeah, cause I moved up from Cornwall in year 4 and then year 5 went to...was it Mundella...to go and see if they had the options and they didn't have the options and the school turned round to me and said you either come to the school now and do whatever, wherever we've got places for or you go back to Cornwall and finish off your education...and that was it. I never went to school again. So I never had a leaving certificate, never took an exam.

Despite having dropped out of school she had managed to get work and bring up her children in a relatively stable environment. This historical account was one of rejection by the system, of not fitting in and of the messiness of moving between schools. She proceeded to describe how she "bludgeons" her way through the mathematical components of her job, and how she has recently had further opportunities to study again but did not complete the course because of family illness. Her family situation meant that there was not enough support available to enable her to fulfil her role as carer as well as completing these courses.

Even though Stacey's mum recognised the negative impact of instability on her own learning, addressing this issue in isolation has evidently not had a transformative effect on her children's success in school. As I described above there are the contributory factors of language and style, as well as the amount and kinds of support that she gives to her children. Another aspect of reproduction indicated by this excerpt is mum's acknowledgement that "I have so many problems." The negative undertones of an invisible oppression seen here are reproduced in Stacey's dispositions as we have seen, but also more precisely. Stacey explained that "my friends at school…I do have problems making them" and in reflecting on how she understood the video camera she described it as "my mum that was zipped up and couldn't speak, so she knew my problems." Talk of having problems is a family trait, and more generally a trait originating in the family's social background. What is also interesting here is the sense of isolation and helpless inevitability about such problems. In addition Stacey reminds us of the fact that she feels that mum doesn't listen, or show concern for the things that worry her.

Understanding this family context is necessary to be able to make sense of Stacey's classroom dispositions and mathematical learning. She was one of a small group of children who had been identified as needing extra support in her mathematics in the run up to the KS2 national tests. The teacher explained that "there's no way that they [a small group of children including Stacey] are going to get level four." Consequently they were in effect denied access to aspects of the class work. However, Stacey was in the classroom for much of the maths lessons that I observed each week. Her lack of confidence was clear in the early part of the lesson when the teacher often used interactive whole-class methods including number cards or white boards. Stacey would often watch other children intently as they scribbled their answers to questions and would then surreptitiously copy them. On other occasions she moved number cards around as they were held up making it difficult to see the numbers. She appeared to use a range of strategies for avoiding the embarrassment of being found to have the wrong solution to a problem. This mathematical anxiety was understood to be a familial condition:

Andy: So Laura's good at art but you don't know what else she's good at. Is there anything that she doesn't like doing?

Stacey: Maths (giggles), it runs in the family!

Andy: It runs in the family, why, who else doesn't like maths?

Stacey: Mum. Dad's okay. Well I don't live with my dad but I see him sometimes.

Andy: So how do you know that your mum doesn't like maths?

Stacey: I asked her to help with my maths. I did some maths and I got all of them wrong and I told her

Andy: And what did she say?

Stacey: It's your work, you should do 'em.

Andy: Do you think that she couldn't do them either?

Stacey: I think she couldn't

Whether or not Stacey's conclusion regarding her mum's mathematical ability was accurate, the effect was the same, Stacey lacked confidence and since she clearly thought (along with her peers) that the teacher's favourite subject was mathematics, she had a problem. On some occasions she was given alternative work to do because she had not understood the previous lesson. So rather than being kept together with the rest of the all-ability class she got left largely to her own devices.

Her mum's response to Stacey's request for help could have been to do with her mathematical ability, as was Stacey's interpretation here, but the impact of this response is also to leave her unaware of what Stacey is doing and whether or not she can complete it successfully. Maybe she expects Stacey to take a lot of responsibility with her schoolwork, unlike Edward's mum who is much more hands-on in supporting Edward with his homework. The effect for Stacey is to allow her to drift towards the disaffected position of her older sister. Lord, Eccles, *et al.* report US studies of the transfer to junior high that indicate that "the most salient predictors of self-esteem change are the adolescents' math ability, physical attractiveness, and peer social skill self-concepts" (Lord, Eccles et al. 1994, 189). Admittedly the context is different but Stacey does not score too highly on any of these areas and so by their reckoning should be expected to struggle. What that research does not explain is the way in which these three important "predictors" are themselves socially constructed.

Edward

The contrast between Edward and Stacey is stark. He is a confident, extrovert child whose parents—now separated—profited from a successful family business. His mum works in a local library as this allows her to organise her time more effectively to look after her children.

> **Edward's mum:** I got this job…it's a job that really fits in with Edward and his life. It's very convenient for me. I don't work in the afternoons I only work mornings, because the main thing I've had in my life ever since I had Edward was that he wasn't going to be a latch-key-kid, and it was really important to me that when I got back from school my Mum was there and I feel that is really important for children so…

This different parenting role is made possible by the economic position of Edward's parents. This not only effects his relationships with his parents and the support that they give him but the cumulative effect upon his development as a learner of mathematics is also clear. When I began working with Edward and his peers they had some idea of what I was researching and were asked to bring something of interest to talk about to our first interview. Edward's choice was his coin collection; gathered from family members' world travels. He attempted to generate some mathematical view of his collection and this willingness to see mathematics in the world around him was a common feature of his interactions with me. I have no evidence to suggest that he was particularly interested in mathematics per se (although he does claim to be) but rather he had a liberal sense of how mathematics was embedded in culture. He seemed very aware of the research game and was quick to elucidate this mathematical view of the world. As well as mathematizing his coin collection he described the school trip to the astronomy research centre Jodrell Bank; of the ratios of planetary sizes and distances; names of shapes; he enjoyed work with coordinates because it was like playing battleships.

He had a masculine view of mathematics and mathematical activity and his mum's beliefs must have influenced this:

> **Edward's mum:** I think he's going to be more of a creative writing kind of person and an artistic person…He used to do some incredible drawings but he doesn't seen to be doing that any more and he just seems to be going more towards the maths and science, which I think's good. I think it's good for a boy. I wouldn't want him

> really to be the arty-farty type; I think to get a good
> career you have to be that way inclined

This perspective on mathematics is also seen in Edward's description of classroom learning. He, like Stacey, is acutely aware of class ranking (although he is not altogether accurate) and speed and competition are closely linked to mathematical success:

Edward's diary: I didn't do too well in the test about maths last time
I only got 35 out of 45 and I was quite disappointed
about that...I think it was the nerves cause I couldn't
really remember doing a proper test before.

[Follow-up interview]

Andy: so were you disappointed because you got 35 out of
45 or because other people got more than you?
Edward: both
Andy: If you had got the top mark with 35 out of 45 would
that have been good?
Edward: Mmm, I would have felt better
Andy: so it's not only about how much you got but who you
did better than?
Edward: yeah I suppose
Andy: who do you want to be as good as?
Edward: I want to be as good as Matt cause today we finished
maths more or less exactly the same time so were al-
ways you know...top

So, like Stacey, Edward has a strong sense of position in the group but his position is near the top and something that he is striving to improve, unlike Stacey who somehow feels powerless to do anything about being "the dumbest kid in the class."

Andy: talking about maths...how would you rate yourself
in terms of the whole class, what position.
Edward: [quickly] third
Andy: after whom?
Edward: Matt, er, well second possibly. I think Francesca would
beat me as well. It depends really as I have never really sat anywhere near Francesca so I don't know how
far she gets. Me and Tim were the first ones to do the
entire thing today cause everyone else didn't do all of
what we did

This idea that speed is related to ability is common amongst these learners (Nicholls 1978; Nicholls 1984) and this view is reinforced by the dominant pedagogic style of the teacher. Mathematics lessons are structured around text-book exercises with rewards and praise usually given for quick and successful completion of work.

Edward apparently studies maths a lot at home, not for the love of the subject but as part of a strategic plan for his future. His Dad gets him to do extra maths work when he is with him on a Thursday as his parents want him to go to city's prestigious fee-paying school for boys. This association is striking as there is recognition in this family of the gate keeping role of mathematics:

> **Andy:** so why do you think that you got into the boys high school?
>
> **Edward:** I was very polite and I had a very long conversation with the headmaster and, erm, apparently the maths teacher could tell that I could see things quite easily without having to work them out. So, I think it was based on maths really
>
> **Andy:** so what do you think are the kinds of people that they are looking for, to go to the high school?
>
> **Edward:** well they're definitely looking for sporty people. They want the majority of sporty people and they were quite impressed with my sport 'cause I do loads and they're looking for music, you know, people that play things like the cello or something. You have got to get over a grade four I think... and then they look at academics as well so it's just... if you get a music scholarship for instance that has got nothing to do with the academics really so I think they just apply for the music scholarship without taking the 11+

The self-perception of being an academic (rather than a musician) is interesting, as is Edward's view that maths played a key role in his securing a place. Also interesting is the role of "manners," those dispositions of the habitus that demonstrate his suitability to be included amongst those with similar privileged backgrounds. Obtaining a bursary-supported place at this private school will cost a lot of money for the family but the investment of economic resources (to be converted to cultural and symbolic capital) is considered money well spent by his mum. One reason for this is the perceived reduction in risk (Beck 1992) that concerns Edward's mother in regard to the normal progression to the local comprehensive school. There he might "go either way" but the investment in the private education will help to avoid him going "completely the wrong way." It is unclear what the right way is but it certainly includes a university education and as far as Edward is concerned (at age 10) this could lead to one of two careers:

> **Edward:** When I'm older I'm hoping to be a palaeontologist... about dinosaurs, so you have to do quite a bit of formula maths with that I think...or open a computer software company where you have to do loads of maths...all the complicated stuff

That he is considering such a career (influenced no doubt by a character from his favourite TV show—*Friends*) is striking enough, but the fact that he can relate this to work that he might be doing now in school is highly unusual. Such employment aspirations and strategic educational vision result from his family context. This is reinforced when Edward's mum was asked about the extent to which she has supported his progress. She describes him as "a self-made man"

> **Edward's mum:** He has really done this all on his own. I never want to be one of these mothers who make them stay in and have extra maths lessons. I didn't want to do that and I don't want to do that now. If at any time Charlie says to me "I'm not happy" then we have to think again. I'm not pushing him that way. I always make sure that he has done his homework and I always make sure that he does his homework in a quiet place.

There are some contradictions here. She does not acknowledge the extent to which her economic and cultural wealth and choice of job have impacted upon Edward; neither does she admit how her support with homework is not natural but the preserve of a certain group of parents (mothers?) who have the time and interest to reinforce the values and priorities of school. Moreover, extra lessons are unnecessary because Edward's dad fulfils this role—not just through using one of the vast array of resources on the market, but by relating mathematics to culture and by learning to think mathematically in a way that might be acceptable to a selective fee-paying school.

This understanding of mathematical power developed through the family is invaluable at this transition point. Although Edward's mum claims to have hated mathematics, she values it (and science) over the more creative, "arty-farty" subjects that she enjoyed and made her career in, despite recognising similar interested in Edward. There are also mixed messages from his dad:

> **Edward:** My Dad absolutely hates maths, when he sees a maths book he goes berserk and runs away from it and my mum hates maths as well. I enjoy maths so I'm the only one really...I've still got a maths book from last year and I've been looking at that and doing sums from it with my Dad and he keeps faking to pass out because he hates maths completely and he's never liked maths even when he was at school he's never enjoyed anything to do with maths.

Yet despite this apparent hatred, the family share a belief that mathematics is useful for future educational and economic success. Mathematics has utility for this family and whether it is enjoyed or not, success is important for access to the family's preferred educational pathway. Although Edward claims to like mathematics the evidence from the classroom is not wholly supportive of this claim. It's clear that he likes being good at it and the sense of distinction that this affords, but he does not have the same level of interest in mathematical thinking as his close friend Matt.

DISCUSSION

School choice is high on the educational agenda in the UK (Gewirtz, Ball et al. 1995; Ball, Bowe et al. 1996; Ball 1997; Ball 2003). Although Edward's parents exercised their right to move Edward to a different school, Stacey went to the same secondary school as the vast majority of her primary school peers. So although there is no doubt a social differentiation effected at the transition from primary to secondary school case study data like this shows how the diffraction of educational trajectories (Noyes 2006) is already taking place in mathematics education and the classroom experiences of these two children has been powerless to seriously reduce this effect. The case studies show that gender is influential in these mathematical trajectories, but as has been shown here in the UK (Connolly 2006), social class is a much stronger correlate of attainment than is ethnicity or gender.

However, concluding this paper without considering how the socially reproductive tendencies of schools and mathematics classrooms can be challenged would be unhelpful. For although Bourdieu's tools offer a convincing theorisation of the way things are (and social stasis was the focus of his work), they are not so useful in generating emancipatory pathways (Giroux 1983). My interest in this paper has been not so much on the education of marginalised groups but on how school mathematics, as part of the wider education system, can act to confirm and/or create the marginalised status of those in society. That is not to say that it always acts to create distinction but rather that the traditional modes of teaching and pedagogical approaches lack the power to address these issues. I was also not as concerned with between school effects as within school and class effect; how one school can help to reinforce the social difference between members of different social groups, despite the ways in which the school apparently hides their social difference through the conformity of uniform, all ability grouping, common curriculum, etc.

The question remains then as to what mathematics educators can and should do to ensure that English mathematics classrooms, and those in other places for that matter, are not complicit in this inequitable process. Firstly there is a need to understand the processes that tend to reproductive probability (not determinism) and hopefully this paper contributes to this understanding. Bourdieu's argument that treating all learners equally is in fact to reinforce the differences between them is an important point in this regard. In Ernest's (1992) account of the establishment of the NC in England referred to earlier he explains how the "public educators" were factored out of the discussion by the powerful political, industrial/commercial and academic groups. These Public Educators

> ...represent a radical reforming tradition, concerned with democracy and social equity... to empower the working classes to participate in the democratic institutions of society, and to share more fully in the prosperity of modern industrial society...

> ...represent radical reformers who see mathematics as a means to empower students: mathematics is to give them the confidence to pose problems, initiate investigations and autonomous projects; to critically examine and question the use of mathematics and statistics in our increasingly mathematized society, combating the mathematical mystification prevalent in the treatment of social and political issues. (p. 36)

So, in response to analyses like this one of how mathematics works in schools, the call must be to the "public educators" to increase their efforts to extend this debate in order to impact mathematics policy and practice. This happens to a degree elsewhere (Frankenstein 2005; Gutstein 2006) and there are some working to these ends in the UK. Unfortunately, the audit and surveillance culture in which we find ourselves makes it difficult to deviate from the mandated curricular pathways and preferred pedagogies (Ball 2003). Harris (1998, 175) explains that:

> ...initiatives currently being imposed on teachers are serving, at one and the same time, to reduce the professional knowledge and critical scholarship which underlies teachers' work, and to decrease the political impact that teachers might bring about through instructional activities.

Teachers need to be educated to recognise the political nature of knowledge, the question not only "whose knowledge is of most value" (Apple 2004) but also "whose pedagogies are of most value." Currently mathematics education in England is in the grip of utilitarian functionalism which has disengaged generations of learners. What is needed is a reengagement with political activism. For example, Lerman (1992) called for the development of a Frierian problem-posing mathematics education; Skovsmose and others (Skovsmose 1994; Skovsmose and Valero 2002) argued for "critical mathematics education" and more recently Gutstein's (2006) has demonstrated how young people can and must learn to read and write the world with mathematics. These are the positions of "public educators." Mathematics education in England is currently not the domain of the "public educators" but unless something changes children like Stacey will continue to be marginalised by school mathematics. On the other hand, those like Edward, who arrive in classrooms with the kinds of mathematical sympathies and family capital that teachers tend to value will be advantaged. His future is apparently secured through merit (particularly in mathematics!), but really the difference between his and Stacey's mathematical trajectory is social distinction.

REFERENCES

Apple, M. (2004). *Ideology and Curriculum - 3rd Edition*. London, Routlege Falmer.

Ball, S. (1997). On the cusp: parents choosing between state and private schools in the UK: action within an economy of symbolic goods. *International Journal of Inclusive Education* 1(1): 1–17.

Ball, S. (2003). *Class Strategies and the Education Market: the middle classes and social advantage*. London, Routledge Falmer.

Ball, S. (2003). The teacher's soul and the terrors of performativity. *Journal of Education Policy* 18(2): 215–228.

Ball, S., R. Bowe, et al. (1996). School choice, social class and distinction: the realization of social advantage in education. *Journal of Education Policy* 11(1): 89–112.

Beck, U. (1992). *Risk Society: towards a new modernity*. London, Sage Publications.

Boaler, J., D. Wiliam, et al. (1998). *Student's experiences of ability grouping - disaffection, polarisation and the construction of failure*. 1st Mathematics Education and Society Conference, Nottingham, England.

Bourdieu, P. (1973). Cultural Reproduction and Social Reproduction. *Knowledge, Education and Cultural Change*. R. Brown. London, Tavistock Publications: 71–112.

Bourdieu, P. (1974). The school as a conservative force: scholastic and cultural inequalities. *Contemporary research in the sociology of education*. J. Egglestone. London, Methuen & Co Ltd: 32–46.

Bourdieu, P. (1998). *Practical Reason*. Cambridge, Polity Press.

Carr, M. and B. Kurtz-Costes (1994). Is being smart everything? The influence of student achievement on teachers' perceptions. *British Journal of Educational Psychology* **64**: 263–276.

Connolly, P. (2006). The effects of social class and ethnicity on gender differences in GCSE attainment: a secondary analysis of the Youth Cohort Study of England and Wales 1997–2001. *British Educational Research Journal* **32**(1): 3–21.

Cooper, B. (1998). Using Bernstein and Bourdieu to understand children's difficulties with "realistic" mathematics testing: an exploratory study. *Qualitative Studies in Education* **11**(4): 511–532.

Cooper, B. and M. Dunne (2000). *Assessing Children's Mathematical Knowledge: social class, sex and problem solving.* Buckingham, Open University Press.

Davis, P. (1993). Applied Mathematics as Social Contract. *Math Worlds: Philosophical and Social Studies of Mathematics and Mathematics Education.* S. Restivo, J. P. V. Bendegum and R. Fischer. New York, State University of New York Press: 182–194.

Ernest, P. (1992). The National Curriculum in Mathematics: Political Perspectives and Implications. *The Social Context of Mathematics Education: Theory and Practice.* S. Lerman and M. Nickson. London, South Bank Press: 33–61.

Frankenstein, M. (2005). Reading the World with Math: goals for a critical mathematical literacy curriculum. *Rethinking Mathematics: teaching social justice by the numbers.* E. Gutstein and B. Peterson. Milwaukee, Rethinking School, Ltd: 19–30.

Gewirtz, S., S. Ball, et al. (1995). *Markets, Choice and Equity in Education.* Buckingham, Open University Press.

Gillborn, D. and D. Youdell (2001). The New IQism: Intelligence, "Ability" and the Rationing of Education. *Sociology of Education Today.* J. Demaine. Basingstoke, Palgrave: 65–99.

Giroux, H. A. (1983). *Theory and Resistance in Education.* London, Heinemann Educational Books.

Gutstein, E. (2006). *Reading and Writing the World with Mathematics: Toward a Pedagogy for Social Justice.* New York, Routledge.

Harris, K. (1998). Mathematics Teachers as Democratic Agents. *Zentralblatt für Didaktik der Mathematik/International Reviews on Mathematics Education* **30**(6): 174–180.

Heymann, H. W. (2003). *Why Teach Mathematics: a focus on general education.* Dordrecht, Kluwer.

Howson, G. and B. Wilson, Eds. (1986). *School Mathematics in the 1990s.* Cambridge, Cambridge University Press.

Lerman, S. (1992). Learning mathematics as a revolutionary activity. *The Social Context of Mathematics Education: Theory and Practice.* S. Lerman and M. Nickson. London, South Bank Press: 170–177.

Lord, S., J. Eccles, et al. (1994). Surviving the Junior High School Transition: Family Processes and Self-Perceptions as Protective and Risk Factors. *Journal of Early Adolescence* **14**(2): 162–199.

Nardi, E. and S. Steward (2003). Is Mathematics T.I.R.E.D? A Profile of Quiet Disaffection in the Secondary Mathematics Classroom. *British Educational Research Journal* **29**(3): 345–367.

Nicholls, J. (1978). The Development of the Concepts of Effort and Ability, Perception of Academic Attainment, and the Understanding that Difficult Tasks Require More Ability. *Child Development* **49**: 800–814.

Nicholls, J. (1984). Achievement Motivation: Conceptions of Ability, Subjective Experience, Task Choice, and Performance. *Psychological Review* **91**(3): 328–346.

Noyes, A. (2004). Video diary: a method for exploring learning dispositions. *Cambridge Journal of Education* **34**(2): 193–210.

Noyes, A. (2006). Distributing newly qualified teachers: considering processes of social harmonisation. *submitted for review*.

Noyes, A. (2006). School Transfer and the Diffraction of Learning Trajectories. *Research Papers in Education* **21**(1): 43–62.

Skovsmose, O. (1994). *Towards a philosophy of critical mathematics education.* Dordrecht, Kluwer Academic Publishers.

Skovsmose, O. (1998). Linking Mathematics Education and Democracy: Citizenship, Mathematical Archeology, Mathemacy and Deliberative Action. *Zentralblatt für Didaktik der Mathematik (International Reviews on Mathematics Education)* **30**(6): 195–203.

Skovsmose, O. and P. Valero (2002). Democratic access to powerful mathematical ideas. *Handbook of International Research in Mathematics Education.* L. English. London, Lawrence Erlbaum: 383–407.

Zevenbergen, R. (2001). Language, social class and underachievement in school mathematics. *Issues in Mathematics Teaching.* P. Gates. London, Routledge-Falmer: 38–50.

Zevenbergen, R. (2005). The construction of a mathematical *habitus*: implications of ability grouping in the middle years. *Journal of Curriculum Studies* **37**(5): 607–619.

CHAPTER 5

SOME TENSIONS IN MATHEMATICS EDUCATION FOR DEMOCRACY[1]

Iben Maj Christiansen
University of KwaZulu-Natal, South Africa

ABSTRACT

In this chapter, I discuss some links between mathematics education and democracy, what these links could imply to what and how we teach, and the issues that arise from trying to further these links. I first suggest three links between mathematics education and democracy formulated on the basis of experiences in Denmark, in particular: learning to relate to authorities' use of mathematics, learning to act in a democracy, and developing a democratic classroom culture. The first two are discussed in relation to narratives from real life, with a focus on the tensions which they reveal. From the discussion following the first narrative, it is clear that what is a competency in one context may not be so in another. This is supported by the second narrative which also questions what is most relevant to students in South Africa and thereby gives rise to the formulation of a fourth connection between democracy and mathematics education, related to issues of access. The third narrative informs a discussion of what it means to be critical. It also continues to address the potential tension between wanting to promote students' critical skills and a

International Perspectives on Social Justice in Mathematics Education, pages 69–86
Copyright © 2008 by Information Age Publishing
69

democratic classroom culture versus wanting to support students in learning what others have developed and what is required in order to succeed in the schooling system. Finally, democracy is linked to the idea of "mündigkeit," or "personal authority." This is not only an issue in relation to the students, but also in relation to teachers. On this basis, I briefly touch on teachers' possibilities for making choices concerning what and how to teach. This comprises a fifth connection between democracy and mathematics education.

MATHEMATICAL COMPETENCIES
FOR DEMOCRATIC PARTICIPATION

Werner Blum and Mogens Niss (1989) and Niss (1987) list five groups of arguments for introducing modelling and applications into the mathematics curriculum. Though not mutually exclusive, they reflect very different goals for education, obviously involving socio-cultural/ideological values (Ernest, 1991; Niss, 1987: 6). The two goals most directly linked to democracy appear to be:

- Promote and qualify a critical orientation in students towards the use (and misuse) of mathematics in extra-mathematical contexts.
- Prepare students to be able to practice applications and modelling—in other teaching subjects; as private individuals or as citizens, at present or in the future; or in their future professions.

The first goal is relevant because of the wide range of contexts in which mathematics is applied, where the purpose of the modelling may vary (descriptive, predictive or prescriptive; cf. Davis & Hersh, 1986/88: 115–121) or the nature of the foundation of the model may vary (from theoretically very strong to consisting of a collection of rather ad hoc assumptions; cf. Emerik, Gottschau, Karpatschof, Møller, & Nørgård, 1981; Jensen, 1980).

> The identity of methods and procedures masks the total diversity of situations and encourages the indiscriminate use of certain non validated mathematical methods in totally unacceptable contexts as well as in acceptable, productive contexts. (Booß-Bavnbek & Pate, 1989: 168)

For an in-depth discussion of types of models, the necessary types of critique and its relevance to mathematics discussion, see (Christiansen, 1996). Included in the second option may be the use of mathematics to create awareness of what for some would be considered problematic societal situations, as probably most strongly demonstrated in the activities for adult learners discussed by Marilyn Frankenstein (Frankenstein, 1981, 1983, 1990). A third obvious link between mathematics education and democracy is the development of a democratic and egalitarian culture in the classroom (cf. Ellsworth, 1989; Young, 1989). As Povey (2003) states it:

> To harness mathematical learning for social justice involves rethinking and reframing mathematics classrooms so that both the relationship between participants and the relationship of the participants to mathematics (as well as the mathematics itself) is changed (p. 56).

For the purpose of this paper, the first two points, which so obviously involve mathematical competencies and using mathematics as a thinking tool, will be my focal point.

First Narrative:
Mathematical Competencies in Critique of Models

A month before Christmas 1999, the summaries of the Danish news contained the following:

CITIZENS CHEATED OF 43 BILLIONS?

A 74 year old pensioner, Hans Peter Scharla Nielsen, today summoned the Ministry of Finance, demanding payment of an amount which the state wrongfully has withheld from him—and in his opinion from all other citizens in Denmark. He has meticulously studied the financial models which the Ministry of Finance uses in calculating the key numbers from which pensions, social security, and unemployment benefits are calculated—as well as the different taxation limits. His claim is that the Ministry of Finance since 1996 systematically has used misleading numbers for the average salaries in Denmark. Instead of looking at all salary earners and their salaries as a basis for calculating the average salary, the Ministry of Finance has—in collaboration with the Danish employers' association which provides the salary information—chosen to disregard IT companies. Also, they do not make corrections to make up for that several employees go from being appointed on a group contract basis to being officials. The lower average implies that the pensions, social security, and unemployment benefits, which according to the law must follow the average, are too low. Hans Peter Shcarla Nielsen thinks that it is an amount around 3 billion Dkr. At the same time, the limits for top and bottom taxation have been set too low—and that implies that the tax payers have paid app. 40 billions Dkr too much!

Hans Peter Scharla Nielsen is not some arbitrary pensioner—he has done it before! In 1996 he obtained judgement that the Ministry of Finance had calculated the pensions incorrectly and paid out 1.5 billion Dkr too much. That lead to a reprimand to the Minister of Finance Mogens Lykketoft (Social Democratic party) and to the reduction of pensions for the following years.

There are many examples that mathematics plays a part in decision processes; cases where it requires a good deal of mathematical competencies to reflect critically on the situation (Blomhøj, 1999). In this case, however, one could claim that it is not what we usually understand as mathematical competencies which make it possible for HPS Nielsen to criticise the existing financial models. He "simply" considered if all information had been utilised "correctly." However, that investigation required the use of numerical values and models, and it required a knowledge that information does not exist in and by itself but is constructed as part of the modelling process—choice of variables, formulation of connections and relations, the determination of constants, etc. It is the same kind of understanding of the choices underlying a modelling process involved in challenging the classification of research with military purposes as "research" rather than as "military spending" (Frankenstein, 1983).

Should we expand our understanding of "mathematics" to include these core modelling competencies and the ability to relate in a critical fashion to models which involve mathematics? In the end, technology has to a large extent made it superfluous to learn the methods and techniques which for so long have dominated most mathematics instruction. But it still takes people to formulate and develop mathematical models and to interpret these as well as to apply and criticise the interpretations. As suggested by Ole Skovsmose (1990), the Ministry of Finance's application of mathematics may well influence who feels competent criticising the decisions from the Ministry—even if it is not the mathematics which makes a difference but the modelling process and its underlying assumptions. In other words, the use of mathematics may exclude someone from (feeling confident) taking part in the discussion. It may also change what we (believe) count(s) as arguments (Christiansen, 1996, 1998; Skovsmose, 1990). It remains an open question what the links between mathematical competencies, a general understanding of the limitations of the modelling process derived from specific experiences, and self-confidence are in determining a person's competencies and willingness to challenge political decisions as Hans Peter Scharla Nielsen did.

In Danish society, the use of mathematical models and scientific investigations play a role in legitimising certain political decisions—discussed by among others Morten Blomhøj (1999) and Peter Kemp (1980). The "expert ideology" makes language games which focus on "correct versus wrong" and "efficient versus ineffecient" more acceptable than language games which focus on "ethical versus unethical" or "esthetically pleasing versus esthetically offensive." This also applies when mathematics is applied to realistic situations in the classroom (Christiansen, 1996, 1997, 1998). But as Paola Valero has pointed out, this is far from the case in many other countries, such as Colombia:

[...] decisions are made based [...] also on personal loyalty [...], political convenience, power of conviction through the use of language or violent, physical imposition. In this political scenario and "rationality," mathematics does not necessarily constitute a formatting power that greatly influences decision making. (Valero, 1999)

Is a focus on giving students competencies with which to critically consider mathematical models and their use really that relevant in the rest of the world? Is this focus relevant in Denmark—or is it more about becoming aware of and critical towards which discourse is the dominating one? Could and should mathematics education contribute to the development of this competency, simply because mathematics often is a tool in the ideology which relies on expert statements?

Double Purposes in a Task: Furthering Societal Awareness and Learning Mathematics as a Thinking Tool

Inspired by among others Marilyn Frankenstein, who again bases her work on Paolo Freire's ideas, I have tried to use examples in mathematics instruction on all levels. Examples which, in my opinion, would encourage students to use mathematics at a "thinking tool" in critically considering a number of issues in society. Examples which at the same time could assist students in developing an understanding of the mathematical tools and how they may be used purposely. One such example was about distribution of land on "races" in South Africa. I used it first in the teaching of student teachers in Copenhagen, with the following task formulation:

The table shows the share of land which "blacks" and "whites" in South Africa owned in 1981.

South Africa is 1.223.410 km²

	Population (millions)	Share of land (percentage)
'Blacks'	18	13
'Whites'	4	87

How would you illustrate this together with your students?

Which impression of the distribution of land do different methods of illustration give?

One can draw pie charts showing the distribution on races of land and population and experience that they look "opposite." This is where many of the Danish student teachers started. One can illustrate it by dividing the classroom space into two parts where one is 13% of the total space, and then distribute 22 students, each representing a million citizens. This would activate students and make it appear very "real" to them. But that gives the misleading image that all the "blacks" live on top of each other. Instead, one can calculate area per person:

There are app. 8700 m² per "black" citizen in South Africa.
There are app. 260.000 m² per "white" citizen in South Africa.

So a "white" in average owns approximately 30 times as much land as a "black" does.

And one can go out to a field and measure out how many soccer fields is equivalent to the land owned be each "black" in average.

But is all land in South Africa owned by someone?

For comparison (1988 data):

There are app. 4300 m² per citizen in the United Kingdom.
There are app. 39000 m² per citizen in the USA.
There are app. 394000 m² per citizen in Canada.

So in the United Kingdom, people live closer together than the "black" South Africans do. And only in Canada is there more land per person than per South Africa "white," in average. Still, these data say nothing about the quality of land owned by "blacks" and "whites" respectively. And we have not discussed differences within these two broad groups in the population. There is much more to discuss!

Which purposes can such an example serve in a mathematics class? It can make students

- Work with changing relative numbers to absolute numbers, and the other way around
- Illustrate numerical data, for instance illustrate a relative distribution in a pie chart
- Consider the message conveyed by different methods of illustration
- Discuss what quantitative information can and cannot tell us
- Obtain a basis for being critical towards others' use of mathematics and quantitative presentations
- Become aware of a political situation
- Challenge their view of all of Africa as poor

- Have their interest for learning mathematics stimulated through experiencing it as a "thinking tool"
- Be actively involved in the teaching-learning situation

I thought that this example would have great potentials in a mathematics instruction (also) directed towards furthering competencies of relevance in a democratic society. I did think that it would be a bit "touchy" to use the example in South Africa. However, critical situations, critical questions and challenges, dilemmas and dialogue are core elements in an instruction with these critical intentions (cf. Vithal, 2003), as they are bound to raise issues which bring these competencies into play. Therefore, I used the example when visiting the former University of Durban-Westville (now merged with other institutions to form University of KwaZulu-Natal).

> **Second Narrative**
>
> In the class I visited, there were both "african" (mostly Zulu) and "indian" students. The "indian" students started on the task, though they did not seem pleased. The "african" students refused to take up the challenge—they could only be talked into starting on another task in the set. After the lesson, I had a conversation with students from both groups about their resistance towards the task.
>
> The answer was pretty clear. Firstly, they did not like to be reminded of their situation. Rather than seeing the data as information which could be used in political argumentation, and the task as an exercise in working with data to support an argumentation, they felt reminded that they were considered to be worth less than "whites." Secondly, they did not find it "good" to spend time on working with this kind of example in school. It would not, they said, give their future learners the mathematical competencies they need in order to do well, to secure themselves an education or a job.

When I thought that the task would be relevant to the students, it rested on the assumption that knowledge can be used in (political) argumentation, and on the assumption that competencies in presenting numerical data are useful in this connection. I have already raised the question whether these assumptions hold true, here or anywhere. But the narrative also raises questions about what is considered relevant by the students. In two ways.

After all, there is no reason to document the extreme inequalities to "black" students—they know these far too well from their own experiences. Frankenstein points out that inequality becomes more apparent by being considered for entire groups of people (cf. Frankenstein, 1990), but here the extent of the inequality is know far beyond what the numbers document. The questions in South Africa at this time are different ones, and I did not come close to the core of the issue, at the same time as I did not give room for the students' primary experiences. (Not to forget, these students probably experienced me as "white" and foreigner, which could have influenced their reading of the situation.)

So we must ask who has the right to define when something is a good example that can support the development of competencies in using mathematics. Often, it is assumed that students and teacher can agree what characterises a "critical example" which it is worth addressing in the instruction. But it may not be so straight forward. For instance, Renuka Vithal (2003) found that the learners in a South African primary school were very critical towards decisions made at the school, while the teachers tried to pull them away from this type of critical examples and towards more general societal issues. Is content directed towards furthering democratic competencies worth anything if the power relations in the classroom remain the same?

Furthermore, in a society strongly dominated by inequality between different population groups, is it fair to want to further the students' critical competencies directed towards changing living conditions, with the risk that they are disadvantaged under existing conditions? A contradiction also pointed out by Paul Ernest as "personal empowerment versus examination success" (1991: 213), yet we must not ignore the ways in which examination success relates to subjective experiences of personal empowerment as well as to an "objective" increase in power within existing structures. (Or are the students wrong—can the two be united? Did the task indeed not *also* involve the mathematical competencies?) This points towards a fourth connection between democracy and mathematics education, often recognised in the literature in equating mathematics with power, namely that mathematics education must make it possible for the less privileged to improve their lives. Because we cannot call it democracy if the differences in living conditions are outrageous!

> To sum up, democracy refers to *formal* conditions concerning the interplay between the institutions of a democracy, *material* conditions concerning distribution of goods and services, *ethical* conditions concerning equality, and finally conditions concerning the *possibility for participation* and re-action. (Skovsmose, 1994: 29)

The ways in which mathematics relates to these aspects may, as shown here, be in conflict with each other when they infuse educational goals.

Therefore: who has, when it comes down to it, the right to influence the purpose and content of education? Does our insistence on these "critical examples" end up being "imposition of emancipation'?

How would the historically advantaged feel if the educational system really came to function on the premises of the historically disadvantaged? If our cultural capital (Bourdieu, 1983/2004) was depreciated overnight? Would we not object to the purpose and content forced upon us—even if claimed to be emancipatory? If I had been to state my views before the

Durban experience, I would have stressed that transformation is not mainly about access; it is about changing values. But whose values are to be furthered? And who am I, who have access (both to and gained via mathematics), to say that it is mainly about values?

Empowerment through Learning "Pure" Mathematics?

Third Narrative

We are in a fourth grade in a school near the centre of Dallas, USA. It is an area where the mortality for young men is very high; where it is dangerous to be after dark because of, among other, gang fights; where drugs are sold on street corners; and where most children come to school without having eaten breakfast. All the learners in this class are so-called African-Americans.

The learners answer my questions and comment on each others' contributions orally or with hand signals. Both hands straight up in the air means "I agree," moving the arms horizontally over each other means "I disagree" and another signal means "I am not sure."

They work with addition of negative numbers. In the first lessons I had with them, they came to a standstill in their attempt to find a solution to

$$5 + \underline{\hspace{1cm}} = 0$$

and that gave me a basis for introducing the additive inverse. Thus the learners were introduced to -5 as the number you add to 5 to get 0. Addition of the negative numbers can then be introduced through a series of tasks and questions. Right now, I have given the learners the task to find the solution to

$$7 + -4 = \underline{\hspace{1cm}}$$

Most of them think either that the solution is 3 or that it is impossible—we can only add -4 and 4, nothing else is allowed. The leading question is whether they can see 4 somewhere. After some probing, they rewrite the expression using $7 = 3 + 4$, making the original equation into

$$3 + 4 + -4 = \underline{\hspace{1cm}}$$

It is easy for the learners to prove that the answer to this must be 3. And as the two statements are equivalent (we show that by drawing arrows from each symbol in one line to the corresponding symbol(s) in the other line) 3 must also be the solution to the original equation.

The lesson is almost over, so I ask the learners to join in praise of each other. Then a girl raises her hand. She tells us that she looked in her brother's maths book. Her brother is one of the few from this environment who has managed to get into college. "You know what? There is a whole chapter on negative numbers, but my brother cannot figure out how to do it."

I praise the students—imagine, they are doing "college mathematics"!

The philosophy behind this American project, called Project SEED, is to give the learners self confidence through mathematics. Through special tasks and leading questions, they are to help each other in reasoning out mathematical rules and connections. That means that they not only master calculation rules which to most US-Americans seem like an almost endless row of meaningless algorithms. They also develop a first understanding of the structure of mathematics and of the mathematical processes, in particular conjecturing, proving and generalising. Thus, the students obtain competencies which are highly valued in their further education, making them more on par with students from more affluent backgrounds. Finally, the high value associated with mathematics mastery in the USA was thought to contribute positively to learners" self image.

Would this be more agreeable to the South African student teachers from Durban?

At no time did Project SEED question a system where learners are to master this type of mathematics, and where applications mainly are what Wiliam (1997) has called "MacGuffins"—"...a plot device primarily intended to motivate the action in a film, and to which relatively little attention is paid." The power relation between learners and teachers were not altered either—except that we unlike the regular teachers were not allowed to hit the learners.

CRITICAL COMPETENCIES AND MATHEMATICAL CREATIVITY—LINKED?

In contrast to Project SEED, we have narratives from classrooms where the learners actively work with developing mathematics from a practical problem, related to their everyday lives or not, and where they work more on their own, with less probing and fewer leading questions. Such narratives we have from all over the world and from various types of settings. For instance, (Beck, Hansen, Jørgensen, Petersen, & Bollerslev, 1999) and (Slammert, 1993) tell how students (in Denmark and South Africa, respectively) have worked with the task of making three red and three green frogs change places in accordance with certain given rules. In both cases, the learners invented a notation which could help them remember what they did and make it object for systematic treatment afterwards.

These classrooms appear to be more democratic in the way learners and teachers interact (though the task was still chosen by the teachers), because there is more equal participation and the possibility of negotiation meaning (or even task). But what does it mean to the development of competencies? Does it give students experiences with developing mathematics, which both assist them in doing well in the existing educational system, and make them capable of relating critically to the use of mathematics and use mathematics actively in their democratic participation? Or does it make mathematics harder for the majority and hinders that they experience the advantages of being familiar with existing algorithms? When is one type of learning preferable to the other?

Let us, however, for a minute assume that most students have intellectual potentials which are rarely realised in schools—for a wide number of reasons not discussed here. How does working with open-ended tasks (or even with project work which includes some elements of problem posing) promote democratic competencies?

If by critical thinking one understand the ability to consider what could be different—go from the actual to the potential—then it is a central ingredient in mathematical creativity. Lebesgue was capable of developing a new integral because he could go from looking at the function "seen from the x-axis" to "seen from the y-axis" (Lebesgue, 1966). Other examples are discussed in (Kitcher, 1984). See also (Mason & Watson, 1999).

In order to develop mathematics more or less on their own, without too many leading questions, the students have to make choices—in contrast to repeating established habits. To make choices necessitates taking a stance, and it necessitates awareness on what could be different, what has been taken for granted so far—a critical approach. Are these competencies— critical awareness, ability to make a stance, ability to make choices, independence—which can be transferred to other situations, or do they remain tied to mathematics if that is the context in which they were developed? And opposite: are they competencies which can be developed in other contexts or through the learning of particular tools/skills, and then applied in learning mathematics *as well as* in participating in society?

Are they competencies which we are to further at any time, or do they only have limited legitimacy? Are these competencies which all students can develop? How do we avoid that focus on these kinds of more demanding competencies in instruction does not in itself end up being discriminating?

If indeed such competencies are to be furthered, how does one go about it? Lionel Slammert writes about "mathematical awareness": "the inner self-sensing ability of the learner to pay attention internally to mathematical information, whether it be in the form of a shape, a principle, a concept, a method, and/or a system" (Slammert, 1993: 118). And through examples, he suggests that instruction can further this mathematical awareness by working with open problems which contain potentials for mathematical creativity, and by the encouragement to direct attention to the feelings and sensations experienced when working mathematically. I would add to this: by challenging the students' approaches or point their awareness to alternatives.

Teachers as Instruments and Agents with Personal Authority

If instruction is to do this, among others, then that means that *teachers* must make it happen. What then is required of teachers if they are to choose suitable problems with the desired potentials, if they are to know when and how to challenge and direct, if they are to know when to direct students towards or tell them about standard algorithms, and so forth?

For teachers to be able to this, to make the classroom dominated by mathematical creativity, they themselves must be creative, both mathematically and pedagogically. They must be capable of thinking in alternatives, both mathematically and pedagogically.

In order to do this, teachers must be aware of mathematical potentials in tasks and student activities. This requires extensive mathematical knowledge. If teachers are to do this, they must understand why the algorithms work; they must be able to solve the same problem in different ways; and they must have a good understanding of how concepts are connected, of mathematical structures (cf. Ma, 1999).

Similarly, teachers must be aware of pedagogical potentials in relation to both the individual student and the class as a whole. This requires substantial amounts of pedagogical knowledge as well as pedagogical content knowledge (i.e., knowledge about how students learn mathematics, about concept development, etc.).

Paradoxically, in order to free attention for all of this—mathematical and pedagogical awareness and creativity—teachers must also be able to handle elements of their teaching as routine—"through algorithm." The same must be true if they are to be aware of pedagogical potentials or potentials in relation to the furthering of democratic competencies.

Fourth Narrative

The new mathematics curriculum in South Africa is formulated as a list of desired outcomes, both general and specific to the different learning areas. In additional, there are assessment criteria for the various outcomes, varying with level of schooling, some guidelines for the organisation of teaching, pedagogics, and much more. Schooling is supposedly free, but the School Governing Board can decide to require payment of school fees, an upper limit to which is determined by the lowest income of school goers' families. The fees are utilised in hiring additional teachers, paying for materials, etc.

I observed a teacher in a school in the Western Cape and conversed with her about her teaching. It was clear that she tried to teach in accordance with the new guidelines—according to a teacher educator, she mainly used the examples used in in-service training courses (person communication with Lynn Rossouw, University of the Western Cape). But the way in which she understood the guidelines, they could not be realised in her everyday practice. And perhaps she was most concerned about meeting what she thought my expectations were. Note I = Iben:

> T: Did you…did I…meet up with, ooh sorry…with with what you are expecting?

A little later I asked about OBE. The teacher commented (referring to a R1 allowance per child she has for this class for a given period of time):

> T: I find it is the OBE's very expensive […] After that we have to buy our own paper
>
> I: So you bought this paper and you bought these smarties [box of small chocolates]
>
> T: Yes so I bought the smarties from their one rand and it's it's about seventy five cents a box
>
> I: Okay
>
> T: You understand so you cannot follow this method all the time

The vital part of democracy is not the possibility of voting as much as what comes before the casting of the vote. There must be real choices connected to making decisions, there must be a possibility to debate options in depth, and there must be awareness on the importance of these elements and thereby a willingness to open up to new perspectives and possibilities; a willingness to change one's mind. Democracy requires "mündigkeit"— personal authority: to have the right to have influence, to be able to speak for one's case, to be responsible for one's actions and agreements. It gives rights but also responsibilities. It puts one under the obligation to engage and to stay informed. It puts one under the obligation to know the actual situation but be open to consider the potentials contained in actuality.

This also holds for teachers in relation to their teaching and the given guidelines for what and how to teach—whereby the connection between democracy and mathematics education is supplied with another level. For teachers to act with personal authority implies being open to considering the potentials in the actual instruction as well as in new curricula, guideline documents, etc. It puts them under the obligation to be well informed and critical both mathematically and concerning pedagogical content knowledge. An obligation to object when a potentially empowering curriculum is countered by limited assessment criteria representing traditional values and hindering mathematical creativity. An obligation to object when teachers' space for pedagogical creativity is limited by recipe-like descriptions of how to teach.

In extension of these obligations, it must also be a democratic right for teachers to have a say in how curricula, guidelines and recommended teaching materials are put together; a right to have the many years of experience from the teaching profession being put to use. A right to be taken serious if they choose to criticise curricula and required teaching methods for being too idealistic and too demanding to realise in practice. Do we secure these rights?

SUMMARY

The first part of the discussion in this chapter was first intended as a message to the Danish mathematics education community: that we had been too ethno-centric in our view on the connections between democracy and mathematics education. Looking outside the country makes us realise the assumptions on which we base our discussions: that perhaps the reference to expert statements matter more some places than others, and forgetting that the inequalities in access to resources and education is rather different in other parts of the world.

Naturally, rethinking these points in the light of my South African experiences changed the focus. I found it necessary to develop the points about equality being important to democracy and about teachers' rights and obligations—both elements which to a large extent are assumed to be in place in Denmark, though it has been questioned by critical voices.

What now stands out is a very inclusive perspective on democracy and mathematics education. In my discussion, I have first considered democratic issues in relation to mathematics education as a whole—what we must teach in order to further democracy. One aspect hereof is to develop relevant competencies, so students can empower themselves to deal with authorities in society. Out of this comes the idea of using critical examples, in particular by including modelling and critical reflection on modelling in the mathematics classroom. This is, however, challenged by the other aspect, namely on furthering equality in access and living conditions. The question remains: how are we to balance these two considerations?

Not pretending to answer this, an additional twist was added by my discussion of mathematics itself, where I have suggested that critical thinking vital to acting in a democracy also play a central part in mathematics—though not in any respect implying that one will automatically lead to the other.

The question of "how to teach" is a general educational issue. I only briefly mentioned the issue of a democratic classroom culture with participation and negotiation, where students empower themselves to deal with authorities in the classroom (fellow students as well as teachers). This has, I believe, been given sufficient attention elsewhere. Instead, I discussed the obligations this puts on the teacher to be pedagogically (as well as mathematically) creative. This revealed that democracy in relation to education is not only students being empowered, but also about teachers' empowerment.

I stressed the dual nature of personal authority, the rights it implies as well as the obligations it puts on you. If students are to exercise their rights in a democracy, they have an obligation to put an effort into it, including developing the necessary competencies through schooling (which needless to say implies a right to have access to appropriate schooling!). If teachers are to exercise their rights in a democracy, they have an obligation to put an effort into it, including developing the necessary competencies to engage in dialogue with authorities (which needless to say implies a right to be heard!). As stressed by Renuka Vithal (2003), authority and democracy are not exclusive, they are complementary—in opposition, yet necessary for each other's existence.

POSTSCRIPT: THE RESEARCHER'S PERSONAL JOURNEY REFLECTED IN THE CHAPTER

The chapter is also a selective biography with narratives from the three countries in which I have lived and worked. My experiences from living in Denmark and engaging with the use of mathematical modelling in our society were reflected in the first narrative. Teaching student teachers in Denmark and South Africa turned out to offer rather different perspectives, and that was the basis for my second narrative. The third narrative served more as a contrast, though it also revealed how much students can learn with strong guidance. It is formulated on the basis of my experiences from more than a decade ago, teaching in Dallas. When I came to South Africa for a longer period for the first time, I got engaged with research on actual practices in schools in the Western Cape. This is the origin of the fourth narrative.

The series of narratives reflects how a researcher's personal experiences and culture, as well as teaching and research activities, are linked, how they limit each other as well as contribute to the development of each other. It reflects how bringing one's experiences together through critical reflection will continue to add new understanding to past events. It weaves a network of meaning through my activities which can only be seen in retrospect. Thereby, it becomes a recreation of my lived narrative of my past and thus of myself, and in that sense it shapes my present and future as scholar, supervisor, lecturer, colleague, citizen and mother.

To me, writing a paper based on such glimpses from my life reflects a deep recognition that research is not an activity which is separated from one's values and from who the researcher is as a person, nor is it an activity which happens in isolation. Writing this paper based on personal experiences is trying to be as good as my word, as I have often argued that reflections on one's own experiences is indeed valuable contributions to the community. I hope the reader will agree with me on this.

NOTE

1. An earlier version of this paper has been presented in South Africa at the 6th conference of the Association for Mathematics Education of South Africa.

ACKNOWLEDGEMENTS

Thank you to Lynn Bowie, University of Cape Town, for pointing out where the paper needed clarification; to Renuka Vithal, University of KwaZulu-Natal, for suggesting me ways in which to categorise my thinking and for inspiration on including the links to my personal journey; and to Lionel Slammert, Cape Technikon, for assisting in making the final touches. Thank you also to learners, students, teachers and colleagues who have shared experiences and discussions with me, thus contributing to the shaping of my ideas—as well as my identity.

REFERENCES

Beck, H. J., Hansen, H. C., Jørgensen, A., Petersen, L. Ø., & Bollerslev, P. (Eds.). (1999). *Matematik i læreruddannelsen: Undersøge, konstruere og argumentere 1.* Copenhagen: Gyldendal Uddannelse.

Blomhøj, M. (1999). Matematikkens rolle i samfundet og dens betydning for almen matematikundervisning. In M. Blomhøj & L. Øhlenschlæger (Eds.), *Rapport fra konferencen "Matematik i samfundet - en begrundelse for matematikundervisning?"* Roskilde, Denmark: Center for forskning i matematiklæring.

Blum, W., & Niss, M. (1989). *Mathematical problem solving, modelling, applications and links to other subjects: State, trends and issues in mathematics instruction* (No. 183). Roskilde, Denmark: Roskilde University Center.

Booß-Bavnbek, B., & Pate, G. (1989). Information technology and mathematical modelling. *The Zentralblatt für Didaktik der Mathematik/International Reviews on Mathematics Education, 5,* 167–175.

Bourdieu, P. (1983/2004). The forms of capital. In S. J. Ball (Ed.), *The Routledge-Falmer Reader in Sociology of Education* (pp. 15–29). London and New York: RoutledgeFalmer.

Christiansen, I. M. (1996). *Mathematical Modelling in High School: From Idea to Practice* (No. R-96-2030). Aalborg, Denmark: Department of Mathematics and Computer Science, Aalborg University.

Christiansen, I. M. (1997). When negotiation of meaning is also negotiation of task: Analysis of the communication in an applied mathematics High School course. *Educational Studies in Mathematics, 34*(1), 1–25.

Christiansen, I. M. (1998). Cross curricular activities within one subject? Case: Modeling ozone depletion in 12th Grade. *The Zentralblatt für Didaktik der Mathematik/International Reviews on Mathematics Education, 30*(2), 22–27.

Davis, P. J., & Hersh, R. (1986/88). *Descartes' Dream: The Wrold According to Mathematics.* London: Penguin Books.

Ellsworth, E. (1989). Why doesn't this feel empowering? Working through the repressive myths of critical pedagogy. *Harvard Educational Review, 59*(3), 297–324.

Emerik, R., Gottschau, A., Karpatschof, B., Møller, S. K., & Nørgård, K. (1981). Hvor ligger Ballerup? *Naturkampen*(21), 27–32.

Ernest, P. (1991). *The Philosophy of Mathematics Education.* London: The Falmer Press.

Frankenstein, M. (1981). A different third R: Radical Math. *Radical Teacher*, 14–18.

Frankenstein, M. (1983). Critical mathematics education: An application of Paulo Freire's epistemology. *Journal of Education, 165*(4), 315–339.

Frankenstein, M. (1990). *Relearning Mathematics: A Different Third R - Radical Maths.* London: Free Association Books.

Jensen, J. H. (1980). Matematiske modeller - vejledning eller vildledning? *Naturkampen*(18), 14–22.

Kemp, P. (1980). Ekspertise som ideologi. In O. Nathan (Ed.), *Tænk og vælg: En debat om naturvidenskab, teknologi og samfund* (pp. 45–59). Copenhagen: Gyldendal, Nordisk Forlag.

Kitcher, P. (1984). *The Nature of Mathematical Knowledge.* Oxford/New York: Oxford University Press.

Lebesgue, H. (1966). *Measure and the Integral.* San Francisco: Holden-Day.

Ma, L. (1999). *Knowing and Teaching Elementary Mathematics: Teachers' Understanding of Fundamental Mathematics in China and the United States.* Mahwah, New Jersey: Lawrence Erlbaum.

Mason, J., & Watson, A. (1999). Getting students to create boundary examples. *TALUM Newsletter*(11).

Niss, M. (1987). *Aims and scope of applications and modelling in mathematics curricula: Manuscript of a plenary lecture delivered at ICTMA 3, Kassel, FRG, 8.–11.9.1987* (No. 145). Roskilde, Denmark: Roskilde University Center.

Povey, H. (2003). Teaching and learning mathematics: Can the concept of citizenship be reclaimed for social justice? In L. Burton (Ed.), *Which Way Social Justice in Mathematics Education?* (Vol. 3, pp. 51–64). Westport, Connecticut: Praeger.

Skovsmose, O. (1990). Reflective knowledge: Its relation to the mathematical modelling process. *International Journal of Mathematical Education in Science and Technology, 21*(5), 765–779.

Skovsmose, O. (1994). *Towards a Philosophy of Critical Mathematics Education* (Vol. 15). Dordrecht: Kluwer Academic Publishers.

Slammert, L. (1993). Mathematical spontaneity: Innovative mathematical learning strategies for South Africa today. In C. Julie, D. Angelis & Z. Davis (Eds.), *Political Dimensions of Mathematics Education 2: Curriculum Reconstruction for Society in Transition* (pp. 115–121). Cape Town: Maskew Miller Longman.

Valero, P. (1999). Deliberative mathematics education for social democratization in Latin America. *The Zentralblatt für Didaktik der Mathematik/International Reviews on Mathematics Education*(1), 20–26.

Vithal, R. (2003). *In Search of a Pedagogy of Conflict and Dialogue for Mathematics Education*. Dordrecht: Kluwer Academic Publishers.

Wiliam, D. (1997). Relevance as MacGuffin in mathematics education. *Chreods*(12).

Young, R. (1989). *A Critical Theory of Education: Habermas and Our Children's Future*. Execter, GB: Harvester Wheatsheaf.

CHAPTER 6

UNDERTAKING AN ARCHAEOLOGICAL DIG IN SEARCH OF PEDAGOGICAL RELAY

Robyn Zevenbergen
Griffith University, Australia

Steve Flavel
Consultant

ABSTRACT

In this chapter we discuss a method through which it becomes possible to identify elements of practice that constitute the learning environments of school mathematics. Through this process it becomes possible to identify those elements of practice that may contribute to the success (or not) of students, particularly those from backgrounds which are traditionally marginalised through school practices.

International Perspectives on Social Justice in Mathematics Education, pages 87–103
Copyright © 2008 by Information Age Publishing

INTRODUCTION

Why is it that some students are more at risk of succeeding in school mathematics than others? And of those who are not successful, who are they and why are they more at risk of not succeeding. Historically and traditionally, success and failure in mathematics have been described within two main discourses. The first is that of innate ability where there is some inherent feature of intelligence that predisposes the student for success. The second ties strongly to work ethic whereby there is potential for less able students to be successful but through considerable 'hard work' and 'determination.' These two discourses dominate perceptions as to why students are successful or not. Yet as critical educators, such as Apple (yr) argued, there is a strong correlation between success and background. Within these discourses, this correlation is not seen as problematic as it supports the hegemonic distribution of capital and resources. It plays into the reproduction of the status quo. In this paper, we seek to challenge this position and argue that the practices through which mathematics is taught and learned position students in particular ways so that the status quo is reproduced. What becomes critical is for teachers and educators to understand the hegemonic practices of school mathematics. Far from being overt, such practices are subtle and coercive which is how they remain below the education radar and remain relatively impervious to change. One way in which the critical examination of practice can occur is through what we call an "archeological dig."

Drawing on Bernstein's notion of the pedagogic relay, we argue that both mathematics and culture are relayed to students through the pedagogical practices adopted in mathematics classrooms. Through the teaching of school mathematics, teachers enculturate students into particular ways of seeing and acting in the social world. Some of this is mathematics, some of it culture—where culture can be seen as mathematics culture but of a middle-class, Western form. Thus for students whose culture is not that of the pedagogic relay, coming to learn mathematics is as much about mathematics as it is about the hegemonic culture being relayed through the school mathematics discourse.

For Bernstein (2000), the pedagogic device regulates pedagogic communication through various rules and structures. He argues that the pedagogic device "constituted the relay or ensemble of rules or procedures via which knowledge (intellectual, practical, expressive, official or local knowledge) is converted into pedagogic communication" (Singh, 2002, p. 573). This is achieved through three rules—*distributive, recontextualising,* and *evaluative* which are interdependent but hierarchical. The distributive rule serves to distribute knowledge (and power) differentially among groups so as to produce different pedagogical identities. The recontextualising rule shifts the pedagogic focus from the original discourse (in this case mathematics) into a new form (for example, everyday discourses such as those found in "real-world problems"). As the work of Cooper and Dunne (Cooper & Dunne, 1999) have poignantly shown this recontextualing rule has implications for who is able to access (decode) the discourse and respond appropriately to questions posed in mathematics examinations. The third principle, evaluative, defines what is seen as the valid acquisition of knowledge.

LEARNING AS A SITUATED ACTIVITY

Learning has often been construed as an individual activity that occurs within the head of the learner. This individualist position has been challenged over recent decades to one which is far more holistic and situated within and across contexts. By reconceptualizing learning to be an activity that arises from tensions between various factors, then a more comprehensive view of the complexity of leaning becomes possible. It has been widely recognised that coming to learn mathematics is as much about the mathematics per se as it is about the culture of mathematics. For students whose culture aligns with the practices of school mathematics, learning is not so difficult. In contrast, for students whose culture is different from that represented in and through mathematics practices and discourses, learning is far more complex since the cultural incongruencies are rarely explicitly acknowledged or even known. This makes it difficult for many educators to make explicit the invisible cultural messages contained within the practices and discourses of school mathematics.

The approach that we advocate in this chapter draws on the epistemological position originating from the work of Vygotsky where learning occurred between an object and subject through the effects of a mediating device. Engestrom (2001) refers to this rudimentary approach to activity theory as first generation activity theory. Engestrom (2001) has developed this individualistic model to a more social model where learning becomes far more connected to the community within which it occurs. Internal contradictions become an integral component of the model and as cultural and social aspects of learning become embedded in the model, a third generation of activity theory becomes needed. The object of this more expanded model of activity theory is to consider learning as an activity which is situated in an "entire activity system in which the learners are engaged" Engestrom, 2001, p. 139). In considering the practical application of this theory to learning within a mathematics classroom, learning at the first level is where students come to acquire the knowledge of what constitutes a correct response in the classroom. The second level is where students come to learn the hidden curriculum of mathematics classrooms—what constitutes appropriate knowledge and how to express that knowledge. This is most often represented in the responses given in examinations where students are assessed on their mathematical knowledge but to be able to respond appropriately requires access to much more than the mathematics per se. The third level of learning comes about when students come to question the hidden assumptions about learning and knowledge so as to construct a much broader understanding of mathematics where the cultural, social and historical biases of the curriculum become known and expressed.

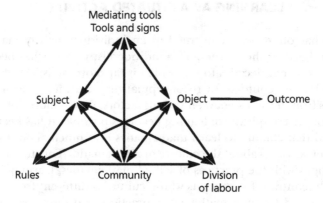

Figure 6.1

What can be interpreted from the model is that the subject (the student) interacts with objects (learner's understandings of mathematics) through the mediating tools (which can be equipment, computers, pedagogies, and the symbolic systems and semiotic systems of schools mathematics). This top triangle is that which is usually representative of the initial work of Vygotsky. However, the model where the lower part of the larger triangle is considered, a much richer theory of learning can be developed. When considering Indigenous learners being placed in school mathematics classrooms, the community of schools and their communities become incorporated into the model. Similarly, the rules of the activity become a focus where issues such as values, the culture of school mathematics, and issues of assessment and other forms of accreditation become incorporated. The final aspect to consider in this larger representation is that of the division of labour where the roles of the participants (learners, teachers, elders) become part of the considerations. Stevenson (2003) describes the elements of the model thus:

> abilities contributing to the enterprise are the *objects* or motives of the collective activity".... The subjects are those in the activity system working together towards this motive for example, the learner, teacher, [students, teacher aides,

and elders]. Together, and with others who share the same common motive, [e.g. education authorities, members of the local region, business people] would make up the *community*. The collective teaching and learning activity is mediate by a large variety of *instruments* (tools) (e.g. equipment, materials, teaching and learning theories...manuals, texts). It is also mediated by rules if they adopted (cultural norms fo the setting), by the ways in which the activity is organized (division of labour) and the community involved in the setting. (249–250)

Each of these elements interact with others as teachers go about their craft. Working in one aspect impacts on the others. If a teacher, for example, decides to develop new resources that embody aspects of the cultural group with whom she is working, then these tools will mediate the teaching and learning processes. In so doing, it may well be that the resources (tools) undergo other transformations as the teacher learns more about the potential of the resources. The new resources may also impact on the rules of the activity. As the resource may require greater interaction among the students (where the interaction had been previously individualistic) then new rules for the activity will emerge.

When working in reform classrooms such as those identified by Boaler, the activity in the classrooms has been radically transformed in a number of the elements of the activity. Teachers designing tasks that embrace higher level thinking and engagement among students, change the rules of interacting within the classroom so that new rules have to be developed. Similarly, the changed expectations of students have created changed circumstances for the activities that are undertaken—students need to engage with different mediating tools—tasks, hands-on activities, performance assessment—which create changes in the objects.

To develop quality learning environments for students who traditionally have been excluded in and through the study of school mathematics, it is recognised that many features of school mathematics work against their success and participation.

ARCHAEOLOGICAL DIG

To uncover the cultural aspects of mathematics, we have coined the term 'archeological dig' as it is much like the digs undertaken in particular sites where the role of the archeologist is to uncover the ways of the world from the remnants left behind by the antecedents. By digging through the remnants left in classrooms, artifacts can reveal much about the culture of the site. Just as the archeologist digs, delves and dusts through obscure items in search for keys to unearth the mysteries of the culture to which the items belonged, so too the process can be adopted for classrooms. We suggest that there is, among all the varied forms of archeologists, is the ethno-archeologist whose task becomes one of searching through classroom artifacts to predict the cultural dimensions of mathematical classrooms.

Using an activity theory approach to understanding the archeology of a classroom enables us to unearth the antiquities of the classroom and in so doing make some observations about the potential learning made possible through that site. In this case, the artifacts that are uncovered through the dig can be seen as evidence of the mediating tools through which the teacher sought to facilitate learning.

Using the metaphor of the classroom being an archeological site, we contend that the principles offered through the archeological tradition enables us to think about the possibility of interpretation of artifacts and how such artifacts were used by the constituents of that community. For example, in a prehistoric dig, how does the archeologist interpret the use of devices that are unearthed, how do these provide insights into the ways of life of the participants. Excavating through the myriad of clues being unearthed, the archeologist continually uncovers clues that enable the construction of life at the time of that culture. The historicity of the site provide clues as to how life was lived at that particular point in time. Similarly, by unearthing the artifacts in a classrooms, the ethno-archeologist is able to make interpretations of the practice of that classroom.

As classrooms vary considerably in how spaces are organized, the use of space becomes a clue as to the approaches used by the teacher. This, in and of itself, provides clues as to the teaching practices and discourses that influence the culture of the classroom. If a simple analysis of classroom and school design are tracked over time, it becomes possible to interpret the mediating tools—in this case theories of teaching and learning—that permeated practice. Small classrooms with no or minimal storage space were common in the early 1900s but many contemporary classrooms now have multiple classrooms connected with folding doors, withdrawal rooms and separate wet/cooking areas. At other times, team teaching was seen as a preferred model of teaching so classrooms were created as large spaces in which teachers shared students and resources. But as digs in these classrooms often show, teachers often 'create' walls to divide the shared spaces so as to create individual classrooms. Such changes in spatial configurations provide clues for the archeological dig as to the views of the participants on how best to organize for learning. Not withstanding the spatial constraint that classroom teachers have little or any control over, there is still some indicators as to how they come to organize that space that provides insights into the learning environment. We recognize that each classroom will be constrained by the physical space so in identifying a systematic approach to the archeological dig, we propose that various aspects of classrooms offer potential sites for collection of information.

Below is a suggested list of what these sub-sites may be and what evidence may be found within these sub-sites.

TABLE 6.1   Potential Artifacts to be Found in a Mathematics Classroom

Site	Artifacts
Teacher's Desk	Student record books, teacher note book, student work, teacher resource books, forms, OHTs, laptop, calculator, whiteboard markers, pens, assessment items, record books, professional journals, syllabus documents
Student Resources	Writing instruments, books for recording work, text books, mathematical equipment (e.g. protractors, rulers, compasses, calculator), laptops, calculators, PDAs, MP3s, mobile phones
Equipment Area	Books, teaching aids and resources, reproducible items such as worksheets, computers and other digital media
Displays	Commercial posters, student work, teacher-made prompts, chalkboard/white board, data projector screen, concrete items (rocks, leaves, dolls, etc)
Classroom resources	Books, Chairs and tables (configurations), computers, laptops, internet connection, chalkboards, flip charts, interactive white boards

In the following sections of the chapter we draw on work we have been undertaking across a number of classrooms to explore the notion of the archeological dig. We present a compilation of a number of classrooms in which we have worked where the classrooms are in traditionally disadvantaged contexts. In presenting these classrooms as an amalgam, our intention is to illustrate the richness and potential of this method.

Triangulation: Keystone to Valid Interpretations

We recognize that it is not possible to collect artifacts and ascribe attributes to them. The archeological dig can only rely on artifacts, there is no scope for interview or follow up. Just as the archeologist can not ask the Mayan how they undertook particular astronomical observations, this must be interpreted from the evidence collected. Where multiple sources of evidence offer similar or same interpretations, a more accurate representation or interpretation can be made of the site. For example, if books were found that focused on the implementation of particular computer technologies and yet no computers were found in the classroom, no digital displays strung across the room, no entries into the teacher's work plan or no evidence of student work in their desks, then it is increasingly likely that the teacher did not engage with this mediating device as a potential for student learning. Conversely, where there was substantial and multiple sources of data that illustrated the use of digital technologies being used to develop, enhance, extend and/or document student learning, then there is a greater chance that this pedagogical tool was an integral part of that classroom. For us, this triangulation between data sources is key to the approach, in that it provides rigour to the analysis and eventual conclusions about the classroom.

MEDIATING TOOLS IN THE DIG SITE

In undertaking a dig within a classroom, the mediating tools are often the first to be uncovered. These tell-tale toolmarks are not only evident in the concrete resources that are unearthed but also in the mediating signs and semiotic systems found in the classroom. The concrete resources include books; teacher planning documents; mathematical and digital resources as discussed in Table One. However, as meaning is made through the cultural message system—that is the discourses used within the classroom, other forms of analysis are also needed. Through careful analysis of resources such as student work samples, student work books, videos, displays in the classroom, it becomes possible to identify the discourses and discursive practices that come to make up the classroom. For example, many Indigenous students come to school speaking their home language which may not be a legitimate one for school language. In some cases, students may speak a number of different languages before they come to school so that the school language is just another language. Multilingualism is common among indigenous students living in remote areas of Australia. How teachers who recognize this diversity of language (and culture) can be identified through the artifacts found in the classroom. For example, some teachers encourage a bilingual classroom that displays both school language and home language signifiers (words) against particular signifieds (concepts) are evidence that the teacher encourages a bilingual classroom where the language and the culture of the learner is valued.

Background to Issues in Indigenous Education

Indigenous students' world views are often very different from that represented in school mathematics. From the outset, the world view of most Indigenous cultures is one of quality and relationships rather than quantity (Watson, 1988). In her work with Yolgnu people of Arnhem Land, Watson (1988) provided very detailed analysis of the complex networks developed by Yolgnu people that provided frameworks on how they saw the world and acted within the social world. Documenting the recursion in these networks she argued that this was a very different organization of the social world from that of the decimal recursion found in Western number systems. Similarly, she documented the ways in which these people also 'sang and signed' the land through cultural and social markers thus creating very different maps than those of Western traditions. In studying time, Harris (1990) documented the qualitative aspects of various people from northern parts of Australia. She shows very different constructs of time and the passing time—where units of time are natural divisions (seasons, night/day); where time is cyclical, and where time is an event. These conceptions of time are in stark contrast to the linear and measured calendars of Western cultures. Such students show the stark differences in how people come to organize and understand their worlds. Such understandings create very different ways of mathematizing the world.

Language is a key aspect of the teaching process. Many Indigenous students coming into Western classrooms are multilingual with school English being a third or other language. Cracking the code of mathematics is as much about learning the language of the subject as it is about the mathematics *per se* (R. Zevenbergen, 2000). When teaching mathematics, teachers need to be aware of the specific language of school mathematics and make this explicit to students in order that they can access the concepts. In observing a very astute teacher of Indigenous students, it was noted that teaching language to multilingual students is a key pedagogical tool (R. Zevenbergen, Mousley, & Sullivan, 2004).Learning takes place within social contexts. The social rules and norms which govern and shape the interactions within these social contexts are bounded by various aspects of culture. Rules of who speaks when shape the discursive interactions of those participating in dialogue. Indigenous people often have oral cultures so that talk is an important part of how they come to see and position themselves. Respect for elders is a key aspect of many indigenous cultures so that the relationships between speakers can be constrained by the rules of status. Similarly, in some indigenous cultures, judgments of learning are not made by the answers given to questions but by the questions that are posed by the learner. Malin (1990) argued that indigenous families encourage independence in their children so that many of the customs adopted in schools (such as asking to go to the toilet) are seen as 'begging' by indigenous people.

For students coming to learn school mathematics requires an explicit recognition that this represents a significant different way of learning but also a very different way of seeing, organising and translating the social world. Western mathematics is premised on a very different set of cultural assumptions than many indigenous cultures. Within Australian Indigenous cultures, there is a strong tendency to see the world within a qualitative framework and where networks between objects/people is fundamental to the social world. It is a most common introduction among indigenous people to find out where a person comes from, who is her/his family so that a set of networks and connections can be developed between the speakers. This is a very different worldview from Western societies where the emphasis on quantity. In these cultures, number is more important –for example, when talking with young children one of the first questions posed is most frequently about the age of the child, how many siblings, their ages and so on. These very different world views significantly shape how the social world is organised so that coming to learn school mathematics for students whose world view is very different from that represented in and through school mathematics, becomes a task of constructing new ways of seeing and viewing the social world. As such, learning school mathematics is not so much about coming to learn numbers and shapes but about a very different orientation towards the world.

A MYTHOLOGICAL EQUITY CLASSROOM

Drawing on our work in classrooms in general and equity sites in particular, we construct a mythological dig on what may be uncovered in a classroom where the teacher was engaged in anti-racist (not the right word) teaching. In this classroom, the teacher drew on many resources to support her teaching and where the object of the lesson was mathematics. The outcome she sought was higher order mathematical thinking for her Indigenous students recognizing that the community wanted their children to be able to walk in both worlds—that of the Indigenous groupings to which they belonged but also have access to white, powerful knowledge. Elements of such classrooms include:

- High achievement
- High expectations
- Oral culture—appropriate communicative practices
- Quality use of technologies—digital and other
- Engagement
- Explicit criteria—students need to know what is expected of them
- Connected learning—connections with the culture of the students
- Integration of culture of students in a genuine partnership
- Recognition of cultural diversity

We are mindful that when conducting the dig, that evidence does not necessarily equate to quality practice. For example, finding a laptop on the teacher's desk does not necessarily mean that the classroom is technologically-rich. Other evidence will confirm or refute the approaches being used. For example, where there is evidence that students have been using computers to foster deep understanding, we would expect to see other sources to confirm deep learning. These could include computers scattered throughout the room, student displays, student work, newspaper clippings where students' contributions have been celebrated, prizes (state, national), etc. Other evidence of quality computer usage could include print outs of a data summary table and analytical report using appropriate software; computer generated representations in both electronic and printed form (graphs, tables, spreadsheets); and computer generated artefacts that support oral presentations (graphs, images, powerpoints). Collectively these artifacts provide the evidence to show that computer technology was an integral aspect of the classroom but also the pedagogical approach used by the teacher. As such, the multiple sources of evidence is key to being able to make valid claims about the approach being used by the teacher.

EVIDENCE FROM A DIG

In considering the dig, it is a collection of evidence at a given point in time. It is reasonable to expect that at any time, teachers focus on particular topics and alongside this topic are pieces of evidence to document student learning; the pedagogy that was adopted; the types of assessment that were used; ways in which the topic was linked to other areas of learning—mathematical, cultural, social. In presenting the data in the following section, we draw on our experiences across a number of classrooms. We do this intentionally to illustrate the approaches taken by a number of teachers as they move towards a socially inclusive classroom. However, we also add, that the data is used to illustrate the approach we are advocating.

In this section we discuss evidence that would be expected when an inclusive approach to teaching mathematics was adopted. Such an approach draws on high expectations that students can and will learn complex mathematics and where there is a recognition and embodiment of the cultural norms of the culture as part of the classroom practices. For many indigenous communities, there is a recognition that the community needs access to Western forms of knowing but this is juxtaposed with the concern that such knowledge can come as a cost to the indigenous culture. As such, many communities seek to adopt an approach where both cultures are part of the approach taken within schools.

Seating Arrangements

Students' desks were organized in small groups of 4—6. At each configuration, central collections of pens and other instruments were provided; cards with roles (leader, recorder, gofer) were placed in a packet on each set of tables. At each collections of tables, was one piece of large, white paper. Student marks (drawings, notes, figures, working out and writing) were on these papers. Writing was in different handwriting suggesting that individual students contributed collectively to the work. There was no evidence to suggest individual recording of work.

Student Constructed Posters

Posters made by the students are hung around the classroom. The topic - percentages—shows a deep knowledge of the concept where students have made links between the various representations. The examples they draw on are ones from their community—demographics of the population, statistics on community issues (wealth/poverty; health, etc). Stories are provided in a narrative and supported with pictures (photographs, drawings).

Student Work

Students have individual crates that contain their work. These are placed to the side of the room as it appears that there are no assigned desks to individual students. The work in these boxes indicates the level of work and presentation of that work. The examples in books and on loose leaves of paper indicate high levels of mathematics being undertaken by the students. In many cases, these are embedded in multiple forms of representations. There does not appear to be any worksheet-type activities in the crates. In most crates, there are computer printouts of work where students have presented their learning in a narrative style and often taken to explain/justify their responses.

Teacher's Planning Documents

The teacher has a large folder in which she has her daily, weekly and term plans. The documents show the expected learning of the students. This is at the expected state levels of achievement and in some cases, exceeds state benchmarks. There is a high level of cultural 'sensitivity' in overt planning in teacher's planning documents. There is explicit recognition of the need to draw on culturally relevant experiences and resources. Notes indicate how she will involve members of the wider community in planning and implementation of rich tasks.

Teacher Resources

On the teacher's desk, there are a number of resource books. There are no schemes but resources that draw on a range of materials. There are some books published by the Western Australian Dept of Education on Indigenous Education (Department for Education of Western Australia, 1999) that outline issues of language and culture in the teaching of Indigenous students; resources created by other teachers in the area and shared through a network; books produced by other sources that outline issues (and statistics) for Indigenous communities; downloads from the web on aspects related to the Bush Medicine Rich Task.

Prompt Sheets

There are two main prompts being used in the classroom. As most students are multilingual and school English is a third or other language, there are many prompts that draw the synergies between the students' language and the mathematical concepts/language. The second prompts are sheets that outline the explicit criteria for work the students are doing. These prompts are useful to the students to enable them to unpack the demands of the tasks but equally important to aides and visitors to the room to access what the students are doing. The use of these resources enable students to be independent in their learning and not have to rely on teacher support for clarification of particular issues that might arise in the lessons.

Assessment

The teacher uses a format of rich tasks where the learning is expected to be transdisciplinary and draw on authentic experiences for the students. The next rich task that is being planned is for the students to plan and build a medicine garden in the school ground. There is substantial mathematics in the task ranging from spatial representations to budget management. There is community consultation and engagement, and the conduct of a survey. Elders will be invited to share their knowledge of bush medicine with the students and to be involved in advising the class on the foods to be planted. Elders will be a key part of the rich task as they will be the source of knowledge upon which the students will need to draw.

Computers

There are a number of computers in the room—some are scattered around the room, and a set of four are placed in one corner of the room. Two seats are with each computer suggesting that students work in pairs at the computer. Prompt sheets at each computer have a list of words (related to bush tucker) and some websites that have been found. These sites are in students' handwriting suggesting that students have found these sites and are sharing their knowledge with their peers. Another prompt sheet (the size of a small poster) is placed behind the computers and shows how to open and enter data on a spreadsheet. Visual prompts are used within this poster along with written instructions.

Physical Space

Room adjoining classroom has been made available for elders to come into the classroom/school. Chairs scattered around the room and tables where elders come and work with students on medicinal plants—drawing on their knowledge.

Biography—Student of the Week

At the front of the class there is a student of the week poster. It contains the biography of student includes photos of family, interests, where he comes from (family). There is a strong emphasis on his (extended) family. The reason for the award is also listed. In this case, the award was given for participating in out-of-school activity (something to do with community).

Class Portrait

A photograph of the class has been blown up on a large piece of paper. Students have drawn arrows to other students and written their relationship to the other person (cousin, sister, etc). Some students have made links that show their family or kinship links to other students.

Class Motto

A large runner has been made with successive pieces of paper joined together and painted with an indigenous design. This runner contains the school motto which represents a celebration of culture and strength—pride in being indigenous and that they can achieve well in school.

Photos of Parent Helpers

To the side of the room is a large pin board that contains photographs of helpers that support the classroom/school. In some cases it is of the person alone, in others they are with children or family. In each case, there is a text that joins the photograph that explains the ways in which the person supports the school. Family and community are strong elements of indigenous people. Recognizing the contribution of family in the school pays respect to those working with the education of their children.

Mathematical Resources

While there is a large area adjoining the classroom which is often used in other contexts for storing materials, the teacher has all mathematical materials and equipment available in the room. These are placed in a section on the side of the classroom and displayed in a way that students easily can see the resources. A prompt sheet is with each box that lists its contents and reminds students to place the equipment back in the box so it will be there for the next student. Students are able to access these resources when they need them so they do not need to ask for permission or need to leave the room.

Attendance Roll

One of the biggest issues in indigenous education is the attendance of students in schools. Many indigenous students have considerable gaps in their learning due to non-attendance in schools. Some of this may be due to cultural issues—such as deaths in families where the family may need to attend a funeral a considerable distance from their home and then remain on to support the family. In other cases, non-attendance may be due to lack of relevance and purpose of Western education to Indigenous families so that non-attendance can be seen as a logical choice to an imposed curriculum. Where schools have turned around the provision of quality education to Indigenous communities, there has been a significant turnaround in attendance. Within an archeological dig, it would be expected that a relevant curriculum that engaged students would result in high attendance.

THE POWER OF THE ARCHAEOLOGICAL DIG

From this mythical dig, it becomes clear that there are a number of features that are reinforced. There are high expectations (mathematical) of the students as evident in the work being presented and the listed expected outcomes in teacher planning. These suggest that the teacher has expectations that the students can and will learn. This is a significant shift from other approaches that focus on deficit and impoverished models of learning. There is strong evidence of cultural recognition and its integral aspect of learning. This is evident in many aspects of this classroom from the students' work, through to teacher planning and the physical layout and resources in the classroom.

The data that can be collected from a classroom provides key insights into the pedagogic device adopted by the teacher/s. In the case presented in this paper, we illustrate the principles identified by Bernstein. In the artifacts identified in this dig we are able to make a number of recommendations. The teachers in these sites were clearly challenging the distributive rule that is common in classrooms working with disadvantaged students. Through the provision of high quality, high expectations and strong scaffolds, the teachers were seeking to shift the relationships of power so that the indigenous students were able to access rich forms of mathematical knowledge and knowing. While we are not able to make comment as to the success of the students in the longer term, the archeological dig provides evidence to suggest the quality of the learning environments in challenging the status quo. The recontextualizing principle was also evident in the ways in which teachers sought to make the mathematics culturally explicit to the students. Not only was the school mathematics a form of recontextualisation for Indigenous students—that is, it represents a very different worldview, teachers sought to make this recontextualisation principle explicit to the students so that they could decode the messages contained within the pedagogic relay. The evaluative principle was evident in the forms of assessment being used by the teachers. Using tasks that enabled the students to represent their forms and ways of knowing mathematically in discourses and practices with which they were highly conversant and in the contexts that were enabling, created opportunities for students to demonstrate their understandings in ways that were culturally relevant and enabling.

REFERENCES

Cooper, B., & Dunne, M. (1999). *Assessing children's mathematical knowledge: Social class, sex and problem solving.* London: Open University Press.

Department for Education of Western Australia. (1999). *Two-Way English: Towards a more user-friendly education for speakers of Aboriginal English.* Perth: Edith Cowan University.

Engestrom, Y. (2001). Expansive learning at work: toward and activity theoretical conceptualisation. *Journal of Education and Work, 14*(1), 133–156.

Harris, P. (1990). *Mathematics in a cultural context: Aboriginal perspectives on space, time and money.* Geelong: Deakin University Press.

Malin, M. (1990). The visibility and invisibility of Aboriginal students in an urban classroom. *Australian Journal of Education, 34*(3), 321–329.

Singh, P. (2002). Pedagogising Knowledge: Bernstein's theory of the pedagogic device. *British Journal of Sociology of Education 23*(4), 571–582.

Stevenson, J. (2003). Integrating approaches to developing vocational expertise. In J. Stevenson (Ed.), *Developing vocational expertise: Principles and issues in vocational education* (pp. 247–265). Sydney: Allen and Unwin.

Watson, H. (1988). Language and mathematics education for Aboriginal-Australian children. *Language and Education, 2*(4), 255–273.

Zevenbergen, R. (2000). "Cracking the Code" of Mathematics: School success as a function of linguistic, social and cultural background. In J. Boaler (Ed.), *Multiple Perspectives on Mathematics Teaching and Learning.* New York: JAI/Ablex.

Zevenbergen, R., Mousley, J., & Sullivan, P. (2004). Disrupting pedagogic relay in mathematics classrooms: Using open-ended Tasks with Indigenous students. *International Journal of Inclusive Education., 8*(4), 391–405.

CHAPTER 7

THE MATHEMATICS CLUB FOR EXCELLENT STUDENTS AS COMMON GROUND FOR BEDOUIN AND OTHER ISRAELI YOUTH

Miriam Amit, Michael N. Fried, Mohammed Abu-Naja
Ben Gurion University of the Negev, Israel

ABSTRACT

The focus of this chapter is an after-school mathematics club, *Kidumatica*, directed towards mathematically talented and mathematically interested middle and early high school students. Since 2002, Bedouin students have been actively encouraged to participate in Kidumatica. The integrative approach adopted by the program has proven successful not only in developing the Bedouin students' mathematical inclinations and skills but also in bringing together Bedouin students with the other Israeli Jewish students in the club in a spirit of camaraderie and with a sense that, through mathematical activity, they stand on common ground. The chapter also suggests that the effect of a circumscribed after-school program, like Kidumatica, which integrates indigenous students with the rest of the student population, may extend beyond

International Perspectives on Social Justice in Mathematics Education, pages 105–126
Copyright © 2008 by Information Age Publishing
105

the students directly involved and ultimately reach the greater community and the schools themselves. In this regard, we believe Kidumatica may provide a model for wider application in other parts of the world.

INTRODUCTION

'Mathematics for All' has been a familiar phrase within the mathematics education community for more than two decades. It has been a watchword for educational policy in the USA since the late 1980s and 1990s (NCTM, 1993, 2000), as it has been in other countries, including Israel (Amit, 1999), which is our concern in this chapter. The meaning and implications of the phrase "Mathematics for All" are not entirely unambiguous (e.g., Fried, 2003; Amit, 2002; Mukhopadhyay & Greer, 2001; Damerow & Westbury, 1985; Keitel, 1987); but whatever else it says, it expresses first of all an obligation, the obligation that no child be denied the materials, conditions, and kinds of teaching necessary for developing good mathematical thinking and the social and economic benefits deriving from it. This is clearly a social obligation, and recognizing it as a central one in mathematics education has caused the discipline to give greater and greater weight to questions of equity, access, and politicization (e.g., Gates, 2006; Tate & Rousseau, 2002; Mellin-Olsen, 1987; Atweh, Forgasz, Nebres, 2001; Skovsmose, O., 1994), beyond the more traditional emphases of problem-solving, learning theories, and teaching in specific content areas. Two main reasons (though there are others) account for this centrality and this shift: one is the realization that students' social, political, economic situations have a tangible influence on precisely those traditional focuses of mathematical education just mentioned (e.g., Lerman, 2000); the other, and the more pressing reason, is the painful awareness that mathematics education, to an unacceptable extent, has *not* reached all students.

Students in lower socio-economic sectors are, not surprisingly, among those who have caused scholars to surmise that the goal of mathematics for all has yet to be truly fulfilled. This is no less true in Israel as it is in other places in the world. But another sector in Israel which has deserved attention under the banner of "Mathematics for All," one which is peculiar to the Israeli case, is that of the Bedouin population in the southern Negev region (Abu-Naja, 2006; Abu-Saad, 1997). Several approaches has been adopted to improve the level of mathematics and science education among Bedouin students. Approaches via teacher education are prominent among these, but the program that we will speak about in this paper goes directly to the students. It is the mathematics club called *Kidumatica for Youth* (hereafter, *Kidumatica*). This program, as we shall describe below, concentrates on developing mathematical thinking in students who show talent and, more importantly, interest in mathematics. But what makes *Kidumatica for Youth* particularly interesting is the way it integrates Bedouins with other Israeli students interested in mathematics; beyond language, religion, and origin, mathematical activity becomes common ground for all the students in the club. In this way, we believe *Kidumatica for Youth* may provide a framework for fulfilling the obligation of "Mathematics for All" with respect to neglected indigenous peoples in other parts of the world.

The paper will proceed as follows. In the first section, we shall provide some general background information concerning the Bedouin population in the Negev, particularly the state of education in their sector. We shall then describe a few different programs and interventions related specifically to mathematics and science education among the Bedouins. A distinction will be made in this section between programs based within the Bedouin community and ones that integrate Bedouin students with other Israeli students. Following this, we shall turn to the *Kidumatica for Youth* program, which is a program of the latter type, and discuss its main features. With that, we shall describe the Bedouin participation in *Kidumatica for Youth*; we shall stress how their participation addresses not only their mathematical development but also their sense of inclusion within the overall population of Israeli students. The concluding section will contain some further thoughts more directly addressing *Kidumatica's* part in social justice, thoughts about the idea of common ground through mathematics, about in-school and out-of-school efforts, and some hope for the future.

THE BEDOUINS OF THE NEGEV: SOCIAL AND EDUCATIONAL BACKGROUND

Although there are also Bedouins in the northern Galilee region of Israel, the majority of Israel's Bedouins inhabit the southern Negev region. The Negev comprises about 12,900 square kilometers; it is a roughly triangular region whose northern side extends from the Gaza Strip to the lower end of the Dead Sea and whose eastern and western sides follow Israel's southeastern and western political frontiers. The Negev Bedouins originated apparently from the Hejaz area of the Arabian peninsula (Ben-David, 1999) and settled in the Negev in three waves between the seventh and seventeenth centuries, each wave displacing the previous resident tribe (Abu-Rabi'a, 2001). Having thus a 1400 year presence in the Negev, the Bedouins may be considered, for all intents and purposes, one of Israel's indigenous peoples.

Today, there are more than 110,000 Bedouins in the Negev (Ben-David, 1999), about half of whom live in seven Bedouin towns established by the Israeli government after 1967. These Bedouin towns were set up initially so that health and other public services could be provided efficiently,[1] but in practice the results have fallen short of expectations: even now, the Bedouin towns exhibit the highest unemployment in the country and suffer poverty and general neglect (Bailey, 1995). It can be argued that these problems (which, it should be pointed out, exist also in the largely Jewish 'development towns' in the Negev, though admittedly to a much lesser degree) arise from the particular nature of the Israeli socio-political fabric and the Bedouins' place in it (Yonnah, *et al.*, 2004); on the other hand, the failures of the Bedouin towns may also reflect, to some extent at least, the same problems typically associated with the urbanization of indigenous peoples all over the world (such as described, for example, in Fischer (1972)).

The last point is quite important, for Bedouin towns mirror the tensions ever at work in encounters with indigenous peoples, namely, those between traditional life and the forces of modernization. These tensions, naturally, are at work also in education. In this connection, the clash between traditional Bedouin community life and modern Israeli life is manifest not only in such things as family size and gender roles (compounding problems already existing with respect to gender in mathematics education) (see Abu-Saad, 1997, p. 32) but in the very notion of formal schooling. For education among the Bedouins was traditionally *informal* and based on the necessities of everyday life. Where education *was* formal, it was the religious *Kuttab* schools, though, during the British Mandate, sons of the tribal sheiks could attend Western style schools (Abu-Saad, 1997, pp. 22–23); formal education, as we think of it, then, was something belonging to the elite. With this background in mind, one can understand how the imposition of modern schools with a typical Western curriculum could pose a threat to traditional Bedouin life and values. Referring to the high dropout rates among Bedouin students, Abu-Saad (1997), accordingly, writes, "Beduin schools have come to represent institutions attempting to diffuse modernization within a traditional community. The emphasis on achievement, as opposed to tribal affiliation and status, is a major revolution in the Beduin ways of life" (p. 33).

The tension between traditional Bedouin life and culture and modern Western-style Israeli life is undoubtedly a component affecting the social and educational state of Bedouins in the Negev. Of course there are other components as well, related to the cultural component but not identical with it: the problem of sufficiently trained and sensitive teachers, just mentioned, is one; appropriate curriculum (Ben-David, 1994; Abu-Saad, 1997) is another; national identity and a sense of alienation (Yonah *et al.*, 2004) may be yet another. We should also mention here the poor physical condition of classrooms and large class size as adverse elements in Bedouin education (Mei-Ami, 2003) (this is a problem in the Arab sector in general, as it is in some parts of the Jewish sector). The educational state of the Bedouins is most likely the complex total result of all these components taken togeth-er rather than any one of them taken in isolation. What is clear, though, is that up to the present the educational state of the Bedouins, relative to the rest of the country, has not been good.

First of all, there is a very high dropout rate among Bedouin students (Ben-David, 1994, Abu-Saad, 1997). Abu-Saad (1997), moreover, points out that girls tend to leave school during the transition from elementary to middle school, and boys during the transition from middle to high school. The high and early dropout rate is one problem which clearly does arise out of the social conditions of the Bedouin community and its cultural background; Abu-Rabi'a (2001) says this quite explicitly: "In brief, the main causes of quitting school amongst the Bedouin are social, cultural, econom-ic, and religious-traditional" (p. 103). Dropping out of school is "a critical problem among the Bedouin..." (Abu-Rabi'a, 2001, p. 99); it is particularly critical for girls, since, accordingly, many are receiving only slightly more than primary schooling—and, for mathematics education, this means they are learning little more than arithmetic.

As for general educational achievement, results on the international as-sessment test, PISA,[2] show an unacceptable gap between Bedouins (as re-flected in the achievement level of the general Arab student population) and the Jewish population, already at the middle school level. And this in-cludes mathematical achievement. Thus, Mei-Ami (2003) points out that, according to the results of the 2002 PISA examination, Israel ranked 31 in mathematical literacy and 30 in reading literacy; however, factoring out the results of the Arab students, Israel's overall rank rises to 12. How well tests such as PISA truly assess mathematical literacy has been questioned (e.g., Jablonka & Gellert, 2001), but the gap in mathematical achievement indi-cated by PISA can be corroborated by a similar gap between the achieve-ments of Bedouin students and other Israeli students on the national ma-triculation examination, the "Bagrut,"[3] taken by students during their high school years.

The Bagrut examinations are prominent both in the Israeli school sys-tem and in Israeli society in general (Amit & Fried, 2002; Amit & Koren, 1995). Success on the Bagrut opens the door to higher education; Bagrut grades are crucial in gaining acceptance to university as well as in determin-ing what majors are open to students after they have entered university. Even jobs and positions in organizations often require one to have taken and passed the Bagrut. And as success or failure in school determines one's own self-view, so does the Bagrut; it is not just another examination (Amit & Fried, 2002). Therefore, despite dramatic improvement over the last fifteen years, the still low success rate on the Bagrut among Bedouin students, rela-tive not only to the Jewish Israeli population but also to the general Arab Israeli and Druze populations, is a significant and worrying fact (Ben-David,

1994; Abu-Saad, 1997; Mei-Ami, 2003; Abu-Naja, 2006). Table 7.1 shows the relative success on the Bagrut ("success" meaning that a student has received a good enough grade on the Bagrut examinations to obtain a Bagrut Certificate, which is roughly equivalent to a high school diploma) among these populations during the ten year period from 1992 through 2002.

TABLE 7.1   Percentages of Students Eligible for the Bagrut Certificate According to Sector

	1992	1993	1994	1995	1996	1997	1998	1999	2000	2001	2002
Total	31.5	32.5	34.0	37.9	38.8	37.7	38.5	41.4	40.8	43.8	46.5
Jews	36.2	37.3	39.5	43.8	45.1	43.7	43.1	45.9	45.6	48.2	51.5
Druze	18.8	22.0	21.4	28.7	27.6	24.1	29.8	35.4	28.6	39.3	36.8
Arabs[a]	19.8	19.8	18.8	22.2	23.1	23.2	27.4	31.5	29.0	33.1	34.0
Negev Bedouins	2.5	3.2	5.1	5.7	6.0	10.3	9.6	13.1	16.8	27.7	25.9

[a] Not including Druze and Bedouins
Source: Shlomo Sabirski, *Eligibility for Bagrut Certificate According to Population 2001–2002,* Adva Center, August 2003).

The possession of a Bagrut Certificate is a requirement for higher education. But it is only a minimum requirement: as remarked above, one must receive a sufficiently high grade for university acceptance. Here too, one finds that the Bedouins lag behind the rest of the population. This can be seen in Table 7.2, which shows the percentages of students who, having passed the Bagrut examinations, have received grades high enough for acceptance to university studies: while in the general population slightly more than 85% of those who pass the Bagrut qualify for university, less than half of the Bedouin students who pass are able to go on to higher education.

TABLE 7.2   Percentages of Bagrut Certificate Holders with Grades Sufficiently High for University Acceptance

	1997	1998	1999	2000	2001	2002
Total	86.2	87.3	86.7	86.2	85.1	85.1
Jews	88.8	89.7	89.2	88.6	87.7	87.3
Arabs[a]	69.4	70.3	69.8	70.4	71.5	73.0
Druze	57.8	65.0	67.0	66.0	66.3	69.1
Bedouins in the Negev	41.2	50.0	47.5	38.4	40.7	46.8

[a] Not including Druze and Bedouins
Source: Shlomo Sabirski, *Eligibility for Bagrut Certificate According to Population 2001–2002,* Adva Center, August 2003).

Actual numbers from Ben Gurion University for the academic year of 2002/2003 can be seen in Table 7.3. Among these Bedouin students who are accepted to university, moreover, one notes that few pursue or are able to pursue the advanced studies in the hard sciences and engineering. This naturally is a particularly painful datum for mathematics educators.

TABLE 7.3 Bedouin Students at Ben-Gurion University of the Negev, by Degree, Gender and Faculty, 2002/2003

| | | Advanced Degree | | Bachelor Degree | |
	Total	Female	Male	Female	Male
Total	319	26	70	116	107
School of Management	11	—	9	—	2
Engineering Sciences	11	—	3	—	8
Health Sciences	40	5	8	14	13
Natural Sciences	16	—	3	5	8
Humanities and Social Sciences	241	21	47	97	76

Source: Abu-Naja, 2006

SOME EFFORTS AND APPROACHES FOR PROMOTING MATHEMATICS EDUCATION AMONG NEGEV BEDOUINS

Having examined some of the main difficulties relevant to the state of education among the Bedouin, we need to turn to efforts that have been made in light of these difficulties, particularly, efforts made to promote mathematics education. The list that follows is not exhaustive and the descriptions of the programs mentioned are brief, but it should give an idea at least of the kinds of approaches that have been adopted. In the next section, we shall consider one of those programs in more detail, namely, the *Kidumatica for Youth* program.

1. "Five-Year Plan"

In 1999, a national plan called the "Five-Year Plan" was announced: its purpose was to promote education among the Arabic-speaking population of Israel and to address the educational gap between it and the rest of the population (Mei-Ami, 2003). It was directed towards the educational infrastructure as well as the academic level of teachers and students. The plan proposed additional study hours, mostly in math and English. As part of the

"Five-Year Plan," an enhancement program was also formed, a systematic evaluation scheme was set out, and a special program for the training of Arabic speaking math teachers was initiated. All these programs naturally included the Bedouin community. In retrospect, however, they were only partly carried out.

2. Teacher Training—Professional Development

One of the central difficulties for the Bedouin community in the Negev, as we have already mentioned, is the lack of good teachers within the community. Like the proverbial chicken-and-egg, the problem begins where it ends. There are few high achievers in the Bedouin high schools, and, of these, only a small percentage of them pursue mathematics and science education. Consequently, there are few teachers to teach mathematics and science; this then creates an over-reliance on teachers from the north, who are far from the Negev Bedouin traditional culture. Moreover, these teachers tend to arrive inexperienced and return to the north before they are able to make a real impact in the Bedouin schools in the south (Ben-David, 1994; Abu-Saad, 1997; Abu-Naja, 2006).

One program addressing this problem, especially for elementary and middle school teachers, was the creation of an in-service mathematics education program for Bedouin teachers. The "Teacher-Professionalization Program," as it was (uninspiredly!) called, was created shortly after the initiation of the "Five-Year-Plan"; it shared the motives and spirit of the "Five-Year-Plan," but it was essentially an independent effort. Once a week, over the course of three years, participating Bedouin teachers studied both mathematics and mathematics education from a theoretical as well as practical point of view. The rationale of the program was that by working over a relatively long term with local teachers, exposing them to a wide range of mathematical and educational issues, a stable and high-quality nucleus of mathematics education in the community could be created and be self-sustaining. The design and rationale of the "Teacher-Professionalization Program" was modeled roughly after another program, *Kidumatica for Teachers*. It is worth mentioning *Kidumatica for Teachers* (described in Fried & Amit (2005) and Amit & Fried (2002)) here, for although it was not specifically for Bedouin teachers, Bedouin participation was actively encouraged. In fact, many Bedouins did participate and studied together with other Israeli teachers as equals; in this sense, *Kidumatica for Teachers* anticipated the integrative approach of its namesake, *Kidumatica for Youth*, which will be discussed below.

3. Upgrading Bedouin Teachers through Advanced Degrees

Ben Gurion University of the Negev, besides being one of Israel's major universities, is the only university in the southern region of the country. Its geographic location in the Negev is part of its identity: serving the Negev region is central to its mission, and, with that, the university has made efforts to encourage Bedouin matriculation into the university and educational development within the Bedouin community. The establishment of the university's Center for Bedouin Studies and Development is a concrete example of such an effort; many of the efforts summarized in this section, however, are also connected one way or another to Ben Gurion University.

A less obvious, but no less important, example of how Ben Gurion University encourages the development of mathematics education in the Bedouin community is the graduate program mathematics, science, and technology education. In general, the program aims to improve the quality of science and mathematics teachers by means of studies towards master and doctoral degrees, while taking into account the previous background of the teacher-student. (the premise-bridging the gap between practice and theory). In recruiting students for the program, a special effort is made to reach Bedouin students. Currently, approximately one third of the students in the program are Bedouin. It is worth mentioning too that the first Bedouin to finish a PhD in mathematics education is a graduate of Ben-Gurion's program; he also happens to be one of the authors of this paper (and his former advisor, another of its authors), which gives testament to the success of the program.

4. "Buds of Science"

Nitaznei Madaa, which means "Buds of Science," was set up in several Bedouin settlements in the Negev with the help of the Center for Bedouin Studies and Development at Ben Gurion University of the Negev. Bedouin pupils from grades 10–11 meet throughout the year with students of natural sciences who come to the Bedouin towns and work with the pupils on aspects of biology, chemistry and physics (the program does not include mathematics). The prime aim is to increase the number of young Bedouins who turn to these subjects after finishing high-school; in the background is the poor achievement of Bedouins on the Bagrut examinations and the meager numbers of Bedouin university students studying the sciences and engineering, which we discussed in the last section. The program has been in existence for about four years, and the level of its success is presently being evaluated.

Another recent program worth mentioning, which was initiated also by the Center for Bedouin Studies and Development, is the "Year of Excellence in the Engineering & the Natural Sciences." Overall, the goals of this program are similar to "Buds of Science," except that it is directly concerned with preparing Bedouins, who have finished high school, for entrance into the university science faculties. It is notable that all books and other material are provided to the students free of charge.

5. "Accessibility to Higher Education"

This program, which has been running for three years now, tries to reach Bedouin high-school students with scientific leanings by bringing them to the university one day a week to enrich their knowledge in natural science, mathematics and technology. The program also helps develop students' knowledge of English—one must not forget that English is the academic *lingua franca*, and, for Bedouin students (and for many other Israeli students), lack of English knowledge is a major impediment to studies in all fields, scientific as well as humanistic. Unlike "Buds of Science," "Accessibility to Higher Education" is not directed exclusively towards Bedouin students: like the teachers in the *Kidumatica for Teachers* and the young people in *Kidumatica for Youth*, Bedouin and Jewish Israeli students study together. It must be emphasized, however, that the mix of Bedouin and Jewish students is an express goal of the program (see Saroussi, 2006). One can discern two basic tendencies in these programs and approaches. Programs such as the *Kidumatica for Teachers* and *Accessibility to Higher Education* bring Bedouin students or teachers into a greater circle of Israeli students and teachers; they improve by focusing on the entire community of mathematics students and teachers and making Bedouin improvement part of the general effort. Thus, it is appropriate to call such efforts *integrative*. Programs such as the "Buds of Science," on the other hand, treat the Bedouins *apart* from other Israeli students, and, here even more so since the activities take place *within* the Bedouin towns. It is right, therefore, to call this kind of effort, *non-integrative*. Integrative and non-integrative approaches, of course, each have obvious advantages and disadvantages. Non-integrative approaches can take into account the Bedouins' particular needs and place them at the center of the effort; however, they may unwittingly further a state of "segmentation" between the Bedouin and Jewish populations in the Negev. Integrative efforts cannot focus single-mindedly on the Bedouins' needs: they will invariably take place in Hebrew, which is the common language, but not the Bedouins' first language; they are not pursued in a setting in which the Bedouins necessarily feel at home. But integrative efforts have the advantages of forming a community of learners and creating a sense of cooperation. These advantages provide strong counterweight to the disadvantages, for they relate to the society as a whole while strengthening each part of it. Let us now turn then to *Kidumatica for Youth*, which is an effort of this integrative type.

THE KIDUMATICA FOR YOUTH PROGRAM

Kidumatica: General Background

Ben Gurion University's *Kidumatica for Youth* (hereafter, *Kidumatica*) was established in 1998 by one of the authors of this paper. Its goal was to create an after-school program in which students from 7th–10th grades[4] could develop their interest in mathematics and their mathematical thinking. Initially, the program was directed towards students in the immediate Beer Sheva area and included about 120 students. Since then it has greatly expanded and now includes almost 400 students.

Once a week, at the university campus, students participate in several "mini-courses," each constructed and led by highly trained mathematics educators. Mini-courses in the past have included: "Logical Problems," "Real-Life Mathematics," "Mathematical Games," "Number Theory," "Algebraic Techniques," "Number Sequences," "Fractals," and "TRIZ" (Theory of Inventive Problem Solving). Once a month, an "activity day" takes the place of the usual courses. During the "activity day," the students take part in mathematics competitions and games, work on the *Kidumatica* newsletter, and listen to special guest lectures on special topics in mathematics, including history of mathematics. It should be underlined that, the lectures aside, all the activities on "activity day" are group activities dependent on high degree of teamwork and cooperation; developing a collaborative attitude in mathematical work is, in fact, one aim of the program—and it has borne fruit in the form of success in national, and even international, "Mathematical Olympiads," which typically demand this kind of teamwork. Most of the students and staff refer to the program as the "Mathematics Club," which further attests to the collegial spirit of *Kidumatica*.

The teaching staff, as we said, consists of highly trained educators: several of the teachers teach as well in middle schools and high schools and some at the technical colleges and university. The majority of the *Kidumatica* teachers are from the former Soviet Union, but not all, and there is now also a Bedouin teacher in teaching staff from one of the Bedouin towns (he quipped recently that his Russian has improved since he joined the staff!). Besides the main teaching staff, moreover, there is a group of upper-level high school students and university students who work as tutors and help with many aspects of the program. Although, during the first few years of *Kidumatica*, these student-tutors came from outside the program, most now are "graduates." This is a very significant side of the program, for, we believe, it makes students aware of their potential for helping other students in the future, that is, for continuing the educative work of *Kidumatica* itself.

To be accepted to the program, students do have to pass a selection examination; however, it must be stressed that the examinations is more a tool for the staff to evaluate the mathematical talent[5] of the entering students than to eliminate students. The truth is, although the great majority of *Kidumatica* students are talented, there *are* weaker students in *Kidumatica* as well. Maintaining the popular character of the program was an extremely important principle in the conception of the program: it aimed to discover and develop the talents of a wide range of students interested in mathematics, not to cater to an elite.

Bedouin Students in Kidumatica

The inclusion of Bedouin pupils in the mathematics club *Kidumatica* has created three important precedents:

1. It is the first program that deals specifically with excellence in mathematics in the Bedouin sector.
2. It is the only mathematics program that actively integrate Bedouin and Jewish pupils.
3. Bedouin girls are present in significantly large numbers (about 40%)

The decision to include Bedouin pupils was made in 2002, specifically, with an eye to create equal opportunities for everyone and with the firm belief that gifted young people among the Bedouins were sure to be found. This belief has been born out: today, four years later, 60 Bedouin boys and girls annually participate in *Kidumatica* on a regular basis, constituting over 15% of the club's members. Two of the Bedouins students compete regularly in *Kidumatica's* top mathematics Olympiad team, and others have taken part in local and state competitions. Most of the pupils come from the Bedouin town of Kseifa and from scattered settlements in the eastern Negev; there are also a number of pupils from the Bedouin towns Lakiya and Tel Sheva. The pupils are from the 8th to 11th grades; all attend high-schools or junior high-schools in Bedouin towns. That approximately 40% of the participants are, at present, girls represents a remarkable, and, we might add, welcome, change in traditional attitudes towards girls' education in Bedouin society.

The Bedouin pupils are quite committed to the program. Statistics collected annually show that Bedouin attendance and participation exceeds that of any other group. Their attendance percentage is 95%, and their dropout rate is 0%. The pupils' commitment to *Kidumatica* is so strong, in fact, that there have been cases where pupils stayed home from school because of illness or to help their parents, and yet, they still came to the club in the afternoon. (This is a contrast to well-documented low attendance in school by the Bedouin pupils in general). The Bedouin attendance and participation statistics are all the more striking in light of the sheer physical difficulty of getting to the program. The students are taken to the university by busses (the transportation is funded in part by the Society for the Promotion of Coexistence and Ben Gurion University's Center for Bedouin Studies and Development, mentioned above), and the trip for some of pupils from the more distant villages can be an hour long. But there is an additional difficulty: many of the pupils live in places inaccessible by bus and must walk nearly an hour to the bus stop. Moreover, because the club is an after-school activity, the equally lengthy trip home means these pupils return to their village or town quite late at night.

Recruitment and Promotion

When the decision first made to extend *Kidumatica* beyond Beer Sheva and to the Bedouin communities, an information campaign was carried out both in schools and among families (we shall return later to the family's role) in order to encourage pupils to attend the selection exams. Since then, *Kidumatica* has become known widely among the Negev Bedouins and there has been much less need for intensive publicity.

At this point, pupils are found mostly through their schools and their teachers' recommendations; but children are also invited to come to the selection exams on their own initiative. In the selection process, every measure is taken to locate true mathematical potential and to eliminate any biases that might arise from language ability or socio-economic background (in Bedouin society there is also more than one socio-economic stratum). To this end, all questionnaires have been carefully translated into Arabic by a mathematics teacher, and Arabic-speaking students are present during examination to answer any questions regarding language or reading comprehension. The questionnaires are also checked afterwards by Arabic-speaking mathematics teachers (the questions are open and often require explanations and justifications).

Language, Culture, and Common Ground

Although the selection examination for *Kidumatica* is given both in Hebrew and in Arabic, Hebrew is the language of the program itself. But mathematics teaching in Bedouin schools, like that of all other subjects, is conducted in Arabic: therefore, even though the Bedouin students learn Hebrew as a second language, it was thought that all mathematical activities and teaching in *Kidumatica* being only in Hebrew might impede the Bedouin students from taking full advantage of the program. For this reason, it was decided bring in, right from the start, two Bedouin university students to work as tutors—eventually, as we described above, Bedouin tutors will be recruited from among the graduates of the program. The Bedouin student-tutors are present in all classes having a large group of Bedouin pupils: when needed, they translate, explain terminology, and provide moral support.

The language difference was certainly a potential focus for tension between the Bedouin and Jewish students. As it turned out, in five interviews conducted with the Bedouin *Kidumatica* students, not one of the students referred to Hebrew as a drawback; and, interestingly enough, two of the students cited the opportunity to improve their Hebrew precisely as one of the attractive and beneficial aspects of the program. They saw Hebrew as a way of improving their chances for success in the university—indeed, they connected the program very much with the university—and they saw this as something to be gained in *Kidumatica* more than in regular school. Thus, in the interview with one of the two students just mentioned, a 9th grade girl whom we shall call here Sana, we had the following short exchange:

Interviewer:	Is there any difference between the mathematics in the Club [*Kidumatica*] and the mathematics in school?
Sana:	There's an enormous difference, really enormous!
Interviewer:	Like what?
Sana:	At school we stick to the books and I feel I am limited. At the Club, there is something different, something new every week—not like school.
Interviewer:	Is there anything particularly special about the Club?
Sana:	Yes.
Interviewer:	Like what?
Sana:	At the university [referring to the Club *as* the university itself] the teachers teach me in the Hebrew language and I think that gives me an opportunity to study both Hebrew and mathematics!

That no tension arose with respect to language might merely be a sign that the Bedouin students recognize the necessity of Hebrew in modern Israeli life; but we believe that it also indicative of the general atmosphere in *Kidumatica* of cultural tolerance and respect. The complete and thorough integration between Bedouin and Jewish boys and girls is one of the distinctions of this project. The Jewish pupils themselves are not a homogenous group: they include new immigrants and natives, religious and non-religious, Ashkenazi and Sephardic.[6] The Bedouins are just part of the fabric: Bedouin girls' wearing traditional clothing and socializing with kids wearing jeans disturbs no one and is taken for granted. The Bedouin pupils participate in all the social activities, including extended day-long activities, competitions, field trips and museums visits. In competitions, the competing teams are typically mixed, Jews and Bedouins against Jews and Bedouins, creating a camaraderie that goes beyond culture and descent. (In the interviews with the Bedouin students, we found only one comment contradicting the sense of equality which *Kidumatica* tries to promote: it was a remark by an 11th grader that vacations are only during the Jewish holidays—a small remark, as the student himself put it, but one which shows there is still work to be done.)

The atmosphere of tolerance in *Kidumatica* may well be, in part, a result of its subject being mathematics and of the students' and student tutors' own views of what mathematics is. They tend to see mathematics as cultureless or in some way transcending culture. Thus, one of the Bedouin tutors told us that "it's easier to work together in mathematics then in other subjects because the arguments are about math and not about everyday things or politics which are more charged…" Another tutor said that "when arguing about mathematics, opinions are not divided according to Jew and Arab, but by mathematical opinion only…" There is no basic position regarding the cultural or cultureless nature of mathematics maintained by the staff and initiators of *Kidumatica*, but there is a recognition that while tolerance means affirming commonality it equally means respecting difference. Thus, when the opportunity arises, special attention is drawn to mathematics from Islamic world as well as problems unique to Bedouin society such as "the camel inheritance problem" which can be related to the complex Islamic inheritance laws.

Family Support

Family is a central pillar of Bedouin life, indeed, of all traditional Arab life. Patai (1976) relates this to the specifically Bedouin concept of kinship spirit or tribal loyalty, a abiyya:

> While a abiyya is thus, in the first place, a Bedouin tribal trait, it was carried over from nomadic to settled Arab society in the form of family and lineage cohesion. Kinship ties, and primarily family bonds, are extremely strong in all sectors of traditional Arab society. They remain an influential factor even after members of a group have moved away from the family home and lived for years in a faraway city or even overseas. (p. 94)

Family involvement and encouragement for the Bedouin *Kidumatica* students is, accordingly, crucial: without it, it is safe to say, the activities would be impossible. At the beginning of each year, we meet with the parents of our Bedouin pupils (it is telling that the meetings are attended only by fathers, though 40% of the pupils are girls). The atmosphere is very supportive and the parents are very proud of their children's inclusion in the club. We hold open discussions at these meetings, mostly in Hebrew, but some also in Arabic. These discussions clearly have shown that the parents themselves place much weight on promoting mathematical excellence. Thus, in one such discussion held in the town Kseifa—which, incidentally, was also attended by the head of the education department, a Sheik and other important members of the community—one father said, "it could lead to a good, profitable profession in the future". Others at the meeting pointed out that studying in the program would make the final exams easier for their kids, and later also help them get into universities, help them get a better start. Comments such as these were made also in conversations with Jewish parents. But there were other comments that related directly to the Bedouin community and Bedouin perceptions of themselves. Many, for example, mentioned that since Bedouin students begin university at a young age (about 3–4 years before Jews),[7] it is important to begin their intellectual development at young age. In a particularly poignant moment, a father said, "Good for you! [addressing the head of the program] Everyone thinks Bedouins are stupid [sic.]: now you understand that there are smart Bedouins, and other people will also see that we have really smart kids..." The parental support demonstrated at these meetings has been corroborated in every interview with the students.

To summarize this section, the primary aim of *Kidumatica* was to provide students talented and interested in mathematics an opportunity to broaden their mathematical experience and develop their mathematical thinking. The commitment to "Mathematics for All" meant that so long as our Bedouin students were not provided with this same opportunity the aim of *Kidumatica* would not truly be fulfilled. The inclusion of the Bedouin students proved successful both in their mathematical attainments and—the aspect we emphasized here—in their total integration with other like-minded Israeli students, with whom they would otherwise have had little contact. The success of the program was given further foundation by the unambiguous encouragement by the pupils' parents and community—this was especially pointed in the case the Bedouin girls. We also ought to mention that integration is never a one way street. Just as the Bedouin students were able to feel their own mathematical power among the other Israeli students and interact with them intellectually and socially, the Israeli students also benefited: many of their stereotypes of Bedouins were challenged, and they could see that, as the father mentioned above put it, "there are some really smart [Bedouin] kids..."

CONCLUDING THOUGHTS

By integrating Bedouin students and other Israeli students into a single community, by having all these students together engage, as colleagues, in mathematical activities, competitions, discussions and collaborative problem-solving, by creating a setting whereby Bedouin boys *and girls* can recognize their own mathematical power, so important in modern life, and a setting whereby other Israeli students can recognize the concrete and potential contributions of their Bedouin neighbors, by allowing friendship between Bedouin and Jewish Israeli students to grow on the common ground of mathematics, *Kidumatica* shows the possibility of a mathematics program that, while it develops mathematical thinking, is also deeply committed to social justice. Its framework shares, in fact, many of the same foundations as democratic societies built on social justice. Hytten (2006) in a piece about education for social justice lists such foundations as follows (referring to the work of Michael Apple and James Beane):

> We create the conditions for a free exchange of ideas, even when these ideas are unpopular, thus allowing us to make fully informed decisions; we have faith in our fellow citizens and in our ability to work collaboratively with them to solve problems and to imagine more enriching possibilities for living together; we employ habits of critical thinking, reflection, and analysis to assess ideas and options, instead of relying on narrow prejudices, uninformed opinions, and personal biases; and we are all concerned with the rights of individuals, the treatment of minorities, the welfare of both intimate and distant others, and, ultimately, the advancement of the common good. (p.221)

That mathematics forms the basis of *Kidumatica* is not, we believe, incidental to the program's success in promoting these democratic values. Whether mathematics is culturally neutral or itself a product of cultural can be argued. Our own view tends in the direction of its being cultural. In the present circumstances, though, this makes it all the more common ground for our Jewish and Islamic Bedouin students, for, historically, western mathematics is what it is largely because of Islamic mathematics. Also applications of mathematics in Jewish and Islamic traditions are both often derived from interpretations of the religious laws of each—for example, in the mathematical-astronomical calculation involved in constructing workable calendars—so that, even where different, a bridge can be found from one culturally related expression of mathematics to the other.

Yet, it is not chiefly those commonalities that make mathematics common ground for our Jewish and Bedouin students, nor is it those that we rely upon in the program itself. Rather, we think, it is that the practice of mathematics, especially mathematical problem-solving (which is so much a part of *Kidumatica*), is such that it demands listening to criticism, recognizing a good idea no matter its source—rich person, poor person, Bedouin, or Jew—working together to get at the bottom of a conundrum, and most of all entertaining the possibility that one may be wrong. Mathematical practice, as scientific practice generally, can be, in this sense, a source of democratic values. That tolerance and social justice are themselves desiderata of mathematics and science was the point of Jacob Bronowski's well-known work, *Science and Human Values* (Bronowski, 1965). There, he wrote:

> The society of scientists must be a democracy. It can keep alive and grow only by a constant tension between dissent and respect; between independence from the views of others and tolerance for them (pp.62–63).

And he went on to say that "In societies where these values [justice, honor, and respect] did not exists, science has had to create them" (p.63). Bronowski may have been overly zealous in his belief, but it can be said at least that the nature mathematical practice is fertile ground for the growth of tolerance and democratic values that we strive to inculcate in *Kidumatica.*

So, it can be said with some confidence that the design of the *Kidumatica* program and the nature of its activities do point to a program that serves social justice. But here we must face a question—and, perhaps, a dilemma. The question is whether *serving* social justice is enough to be a true *vehicle* for social justice? The dilemma lies behind the question and has to do with the obvious fact that *Kidumatica*, as well as several of the other programs mentioned above, is a program taking place *outside* the usual school setting; indeed, it is nearly inconceivable *in* the school setting. On the other hand, school is the central public institution for education: what happens in the schools affects the whole population of Bedouin children; what happens in *Kidumatica* affects, on the face of it, less than a hundred. And school is certainly the locus for most of the educational problems described in the first section of this paper. But, as we said, the kind of activities and framework that make *Kidumatica* successful are hardly possible to implement in the schools, while solutions that are plausible for the schools generally involve great expense and more time than we can afford—even where budgetary allowances are made, the distribution of resources thus made available is a complex and problematic process (Adler, 2001). So the dilemma put baldly, and perhaps over-simply, is between a program that accomplishes much, but reaches few, and schools that reach many, but have limited or problematic options.

The dilemma is really that arising from the conflicting advantages and disadvantages of integrative and non-integrative approaches, described at the end of the second section. Of course, one might say, legitimately, that the dilemma is not a true dilemma since both approaches can be adopted simultaneously—and, ideally, should. We do not dispute this. But, in practice, especially when funding is limited, one must often choose or at least lend one's support to one kind of program or the other. And in this regard, we believe a good case can be made for an integrative program, like *Kidumatica*.

The advantages of the program for the Bedouin and Israeli *Kidumatica* students themselves, we hope have been made clear enough. Naturally, Kidumatica's small budget relative to that of a large scale school program is another obvious advantage But what makes us believe that a program like *Kidumatica* can actually be a *vehicle* for social justice, and not just be related to it in some very limited fashion, begins with the conversations with Bedouin parents described above. In our account, we remarked that at the meeting at Kseifa not only the fathers of the Bedouin students were present but also a Sheik and other central figures in the community: the pride and encouragement expressed by the parents of the *Kidumatica* children were also the pride and encouragement of the community as a whole. In an interview with a 9th grade girl, whom we shall call Nuha, we asked what her friends thought about her participating in *Kidumatica*. She replied: "When I speak about the Club with my friends they get interested and hope they can join too." The point is that the effect of *Kidumatica* does not remain among the participating students themselves. It extends to friends at school and the community as a whole. In effect, the participants in a program, in a club, like *Kidumatica*, belong to *two* communities and thus they become like emissaries from one to the other. This, then, is a solution to the dilemma: an integrative program like *Kidumatica* reaches few students only at one level; by way of those students it reaches many more and the schools themselves. If, ultimately, the students return to their home communities as teachers, then, that would be a true consummation of the process. But even without that, we can see that the positive effects of a program like *Kidumatica* will not necessarily stayed locked within it: the discovery of one's own mathematical power, the recognition that collaboration can bring genuine results, the sense that regardless of whether one is a Bedouin or a Jew one can be worthy of respect—and friendship. It is the possibility that these good effects can spread beyond *Kidumatica* gives us hope for the future.

NOTES

1. Settlement of the Bedouins has also been an issue, and no less problematic, in Arab countries (e.g., Barakat, 1993, pp. 53–54). We might remark that even in Ottoman times, the mutasarrif of Jerusalem Ekrem Bay also proposed the settlement of Bedouins and the registering of their lands, "as means of enhancing stability in the Negev (Abu-Rabi'a, 2001, pp. 13–14), but this, as well as earlier Ottoman attempts at settlement, were never really put fully into effect.

2. PISA (Program for International Student Assessment) is an examination given to 15 year-olds every three years by the Organization for Economic Cooperation and Development (OECD). The number of participating countries in 2003 was 41 and in 2006 was 58. The express goal of the PISA examination is to assess reading, mathematical, and scientific literacy.

3. The 'Bagrut' include examinations in mathematics, English, Bible, history, literature, Hebrew language, citizenship, geography, physics, biology, and chemistry, among others. Of these, mathematics, English, history, Hebrew literature and language, civics, and Bible (Koran for Moslem students) are required to obtain a "Bagrut Certificate," the possession of which is, among other things, a minimum requirement for acceptance into universities. University acceptance depends also on the level of the examinations and the particular grades the student has actually received. Of course, a student who wishes to study a particular field in university may be required by the university to take additional examinations related to that field.

4. There are now some students in the 11th grade as well.

5. More on the identification of mathematical potential in the context of *Kidumatica* can be found in Neria & Amit (2006).

6. Roughly speaking, Ashkenazi Jews are those with European backgrounds, while the Sephardic Jews are those with backgrounds in North Africa, the Middle Eastern Arab countries, and Iran.

7. Jewish boys and girls are obliged after their 18th birthday to serve in the army (2 years for girls and 3 years for boys) or, in some cases, do non-military national service. Bedouins may volunteer for the army, and among the Bedouins in the north of the country, many do: Negev Bedouins, however, generally do not. Therefore, Bedouins that attend university typically begin earlier than the Jewish students.

REFERENCES

Abu-Naja, M. (2006). Bedouins in the Negev: Population, Employment, Education. Talk given in the discussion group on indigenous peoples at the 30th Conference of the International Group for the Psychology of Mathematics Education (PME 30), Prague, Czech Republic.

Abu-Rabi'a, A., (2001). *A Bedouin Century: Education and Development among the Negev Tribes in the Twentieth Century.* New York: Berghahn Books.

Abu-Saad, I.(1997). The Education of Israel's Negev Beduin: Background and Prospects. *Israel Studies, 2*(2), 21–39.

Adler, J. (2001). Resourcing Practice and Equity: A Dual Challenge for Mathematics Education.In B. Atweh, H. Forgasz, & B. Nebres (Eds.), *Sociocultural Research on Mathematics Education* (pp. 185–200). Mahwah, NJ: Erlbaum.

Yonah, Y., Abu-Saad, I., Kaplan, A. (2004). De-Arabization of the Bedouin: A Study of an Inevitable Failure. *Interchange,* 35(4), 387–406.

Amit, M. (1999). Mathematics for All: Millennial Vision or Feasible Reality? In Usiskin, Z. (Ed.) *Developments in School Mathematics Education Around the World* (pp. 23–35). Reston, VA: NCTM Press.

Amit, M. (2002). Equity and Assessment: Challenging the Misassessment of Mathematical Talent in Young Girls. In Bazzini, L., Inchley, C. W., (Eds.), *Mathematical Literacy in the Digital Era* (pp. 269–274). Milan: Ghisetti e Corvi Editori,.

Amit, M. & Fried, M. N., (2002). High-Stake Assessment as a Tool for Promoting Mathematical Literacy and the Democratization of Mathematics Education. *Journal of Mathematical Behavior, 21,* 499–514.

Amit, M. & Fried, M. N. (2002). Curriculum Reforms in Mathematics Education. In Lyn English (Ed.), *Handbook of International Research in Mathematics Education* (pp. 355–382). Mahwah, NJ: Erlbaum.

Amit, M. & Koren, M. (1995). *National Curriculum for Secondary Schools,* Jerusalem: Ministry of Education, Culture and Sport.

Bailey, C. (1995). Dispossessed of the Desert. *The Jerusalem Report,* 26, 64.

Barakat, H. (1993). *The Arab World: Society, Culture, and State.* Berkeley, CA: University of California Press.

Ben-David, Y (1999). *The Bedouin in Israel.* Israeli Ministry of Foreign Affairs. Available at the website, http://www.mfa.gov.il

Ben-David, Y. (1994). *The Bedouin Educational System in the Negev: The Reality and the Need for Advancement.* Jerusalem: Floersheimer Institute for Policy Studies.

Bronowski, J. (1965). *Science and Human Values.* New York: Harper & Row, Publishers.

Damerow, P. and Westbury, I. (1985). Mathematics for all—Problems and implications. *Journal of Curriculum Studies, 17*(2), 175–184.

Davis, R. B. (1989). The culture of mathematics and the culture of schools. *Journal of Mathematical Behavior,* 8, 143–160.

Fischer, H. J. (1972). *Problems of Urbanisation. Bombay: Leslie Sawhny Programme of Training for Democracy,* Friedrich-Naumann-Stiftung

Fried, M. N. (2003). Mathematics For All as Humanistic Mathematics. *Humanistic Mathematics Network Journal,* 27. http://www2.hmc.edu/www_common/hmnj/index.html

Fried, M. N. & Amit, M. (2005). A Spiral Task as a Model for In-Service Teacher Education. *Journal of Mathematics Teacher Education,* 8(5), 419–436.

Glaubman R. & Katz, Y. (2003). The Bedouin Community in the Israeli Negev: Educational and Community Characteristics. In Iram, Y. & Wahrman, H. (Eds.), *Education of Minorities and Peace Education in Pluralistic Societies* (pp. 181–212). Westport, CT: Praeger Publishers.

Hytten, K. (2006). Education for Social Justice: Provocations and Challenges. *Educational Theory,* 56(2), 221–236.

Jablonka, E. & Gellert, U. (2001). Defining Mathematical Literacy for International Student Assessment. In L. Bazzini & C. Whybrow Inchley (Eds.) *Mathematical Literacy in the Digital Era (Proceedings of the 53rd conference of the CIEAEM)* (pp. 119–123). Milan: Ghisetti e Corvi Editori.

Keitel, C. (1987). What are the goals of mathematics for all? *Journal of Curriculum Studies,* 19(5), 393–407.

Lerman, S. (2000). The Social Turn in Mathematics Education Research. In J. Boaler (Ed.), *Multiple Perspectives on Mathematics Teaching and Learning* (pp. 19–44). Westport, CT: Ablex.

McKnight, C.C., F.J. Crosswhite, J.A. Dossey, E. Kifer, J.O. Swafford, K.J. Travers, and T.J. Cooney (1989). *The Underachieving Curriculum.* Champaign, IL: Stripes Publishing.

Mei-Ami, N. (2003). *Background Paper on Academic Achievement in the Arab Sector* (Hebrew). Jerusalem: Knesset–Center for Information and Research

Mukhopadhyay, S. & Greer, B. (2001). Mathematics for All: Rhetoric or Right? In L. Bazzini, C. Whybrow Inchley, (eds.), *Mathematical Literacy in the Digital Era (Proceedings of the 53rd conference of the CIEAEM)* (pp.124–128). Milan: Ghisetti e Corvi Editori

NCTM. (1993). *Curriculum and Evaluation Standards for School Mathematics.* Reston, VA: NCTM.

Neria, D. and Amit, M. (2006) When the Wrong Answer is the "Good" Answer: Problem-Solving as a Means for Identifying Mathematical Promise. In Novotná, J., Moraová, H., Krátká, M., & Stehliková, N. (Eds.) *The 30th Conference of the International Group for the Psychology of Mathematics Education (PME),* 4, 225–232. Prague, Czech Republic.

Patai, R. (1976). *The Arab Mind.* New York: Charles Scribner's Sons

Sabirski, S. (2003). *Eligibility for Bagrut Certificate According to Population—2001–2002.* Tel Aviv: Adva Center.

Saroussi, V. (2006). *Higher Education for Bedouin: The University Program for the Promotion of Accessibility to Higher Education for Negev Pupils 2005/2006 Academic Year.* Available at the website: http://www.kas.de/proj/home/home/24/2/webseite_id-3067/index.html.

CHAPTER 8

SOME THOUGHTS ON PASSIVE RESISTANCE TO LEARNING

Tod L. Shockey
University of Maine, USA

Ravin Gustafson

ABSTRACT

The education community knows that improvements can and must be made for the mathematics education of underrepresented groups. Native American schools, in particular, have struggled due to colonialism, racism, and the mistaken notion of members of the dominant society that Native people wish to be assimilated. In fact, sovereignty is a huge issue for Native people across the country; it is an especially sore point in our state, where a 1981 land claims settlement act clouded rather than clarified Native sovereignty. Mathematics education has learned a tremendous amount through the pioneering academic work done in ethnomathematics. We also know that so many others are doing the good work daily, impacting children's lives, but never receive recognition beyond the walls of their classrooms. We are learning about the positive difference of a Culturally Appropriate Curriculum and finally, we are learning how to respect Native America. This is a preliminary paper that in-

International Perspectives on Social Justice in Mathematics Education, pages 127–137
Copyright © 2008 by Information Age Publishing
127

tends to open a discussion on the challenge of passive resistance to education in a Native American school in the eastern United States.

INTRODUCTION

Mathematics education in "our" reservation school, we hope, is making strides toward improvement. Despite the fact that we failed to meet adequate yearly progress in middle school mathematics for the 2005–2006 academic year according to our standardized state assessment test, we feel that an optimistic look to our future is not unwarranted. We are "getting real" (Kitchen, 2003) about mathematics education reform with a new curriculum, an increase in the number of highly qualified middle school math instructors, a strong focus on mathematics across all content areas and at all grade levels, and a continuing partnership between the school and the state university's mathematics department. The voices in this article are from the "outside," a university mathematics educator, and from the "inside," a classroom teacher.

LOOKING IN FROM THE OUTSIDE

This school's Comprehensive School Reform Plan (CSRP) included the support of numerous consultants, one being the university mathematics educator co-authoring this paper. During the 2005–2006 academic year (this work will continue at least through the 2006–2007 academic year) , the co-author spent two days per week at the school visiting classrooms, observing instruction, acting as a resource and consultant for teachers, attending faculty/staff meetings, and helping organize the first-ever Math Night for the community. He taught a mathematics course for university credit, on-site at the school, during each of the semesters.

TRUST

But if the fieldworker expects to engage in some variety of participant observation, to develop and maintain long-term relationships, to do a study that involves the enlargement of his own understanding, the best thing he can do is relax and remember that most sensible people do not believe what a stranger tells them. In the long run, his host will judge and trust him, not because of what he says about himself or about his research, but by the style in which he lives and acts, by the way in which he treats them. In a somewhat shorter run, they well accept or tolerate him because some relative, friend, or person they respect has recommend him to them. (Wax, 1971, p: 365)

I gained entry to this tribal school through someone I knew: a university colleague who introduced me to the school's curriculum coordinator. The coordinator then took the "chance" of introducing me to the school. I am well aware of the history of research in indigenous settings, and I understand what Smith was referring to when she wrote about research: "when mentioned in many indigenous contexts, it stirs up silence, it conjures up bad memories, it raises a smile that is knowing and distrustful" (p. 1).

As the outsider, I realize as Lincoln and Guba stated in 1985 that trust is "something to be worked on day to day. Moreover, trust is not established once and for all; it is fragile, and even trust that has been a long time building can be destroyed overnight in the face of an ill-advised action." (p. 257). I am keenly aware that others who preceded me in this particular school are either welcome to return or are not. The list of individuals who are not welcome is well understood within the school's leadership and faculty ranks, and there is no hesitation to name names and give reasons. The co-authoring of this article is but one artifact offered as evidence that trust exists and continues to develop.

SCHOOL LEADERSHIP

School leadership has taken on many forms in the past. For a period of about five years there was a new principal each year. Folklore within the school suggests a complete lack of leadership with many of the faculty operating as autonomous units for several years. This autonomy extended to decisions regarding curriculum, student failure/success, and discipline. The school has leadership now that has been in place for over two years, and change is obvious. There is no resistance to reform on the part of the leadership team. External funding is actively pursued through the district office, relations with the Bureau of Indian Affairs exist positively, and decisions are made based on the effect/affect of the students. The leadership team has championed professional development opportunities for faculty and staff within the school but has also opened their doors for teachers from surrounding communities to participate.

In the fall of 2005 a mathematics topics course in algebra was offered on site at the school. For many in the course, this was the first mathematics content for decades. Let's face it, many in-service teachers do not return to University for more credits in mathematics or to pursue graduate programs in the content area. As a mathematics educator teaching this course, focus was on the content, but secondarily we as a class considered what the expectations for our students were, based on the Maine Learning Results. We had rich conversations about pedagogy and curriculum. Many of the teachers highlighted requirements and expectations of the students in this Native school that did not fit the children's reality. As an example a released item from the Maine Education Assessment dealt with dog pens, on the reserve family dogs do not live in pens. Generally speaking, the animal is secured to a lead that is attached to some fixed structure. How, then, were students to make sense of the concept of a "dog pen" when such a thing does not exist in their world view? Another released item focused on comparative shopping, looking for the "best deal." There are two stores within twenty miles of the reservation. When shopping is done, purchases are based on need, not on comparisons. The nearest city that has multiple shopping opportunities is two and a half hours away.

The second semester course focused on geometry. Since our state has a laptop initiative (meaning that all seventh and eighth graders have laptops available to them during the school day), the school purchased a site license for Geometer's Sketchpad®, and that became the delivery mode for the course. We explored many new ideas in Euclidean geometry afforded through this dynamic software. Worth noting here is who enrolled in this course: the school secretary, the kindergarten teacher, the guidance counselor, the fourth grade teacher, a special education teacher, and the assistant principal, as well as all of the middle school teachers. As members of the course reached a comfort zone and could use the program with relative ease, which occurred about four weeks into the semester, the software was loaded onto student laptops. This learning tool is now being woven into the middle school mathematics curriculum, and the broad range of faculty and staff who took the course means that any child with a laptop can seek assistance from almost any location in the building.

The culmination of the course was the first-ever Math Night at the school, and the entire community was invited. With the aid of class participants and the co-author, students, aunts and uncles, parents, grandparents, and siblings explored translations in transformational geometry, creating unique tessellating shapes on a sheet of newsprint. The newsprint was then colored with fabric crayons, and the colored sheet was placed on a T-shirt and ironed. The results were original mathematical pieces of wearable artwork. "Can we have Math Night again?" was heard a number of times from students. We were not expecting so much enthusiasm for mathematics; everyone went home that night feeling good about the excitement of the children!

LOOKING IN FROM THE INSIDE

For the past two years, we have had a stable administration that has done much more than stress professional development; it has provided us with opportunities. During the 2005–2006 academic year, the co-author of this article was available two days a week as a mathematics consultant and one night a week as the instructor of college-level mathematics courses. Teachers and staff at all levels have taken these mathematics courses, and our "final project"—Math Night—was a resounding success.

Native Americans across the country have learned to regard outsiders who come onto reservations to "study" them with distrust. It is no different on our reservation, which is small, has a state highway running smack through the center of it, and is close to a tiny city that has historically been a repository of bad feelings toward its Native American neighbors. My co-author did not swagger onto our campus and tell us what we needed to do for our students. Instead, he sat quietly in our classrooms and watched, ate lunch with the kids, and listened to what we had to say. He was clearly more interested in helping us than in trying to get us to conform to an outside standard of what instruction should be. This is, in every sense of the word, a community school, and my co-author's experience with Native students and his understanding of Native culture have endeared him to some of the elders who come to our school to sit in the office and help answer phones, have coffee with the staff, and eat lunch in the cafeteria with the children. His easy manner and obvious enthusiasm for being in the school have allowed a real sense of trust to develop between him and the faculty and staff. And the kids just adore him.

Throughout the year, our students have consistently been both amazed and amused to see their teachers and other staff struggle with "homework" just as they do. They are seeing us in a new light: we are no longer merely those adults who talk at them all day; we are learners, too. This has opened up many interesting dialogues in the classroom, giving us valuable insights into how our students view their own education. As this interaction continues, we hope to listen carefully and give our children the skills they must have if they are to succeed. We are certainly more aware of our own needs as learners—and of our own responses to and struggles with content.

As much as our skills as educators are improving, though, we still face the challenges of teaching at a reservation school. Our students are well aware that they are surrounded by a society that neither shares their cultural identity nor particularly values their heritage. They do not see themselves reflected in national media; they do not hear voices like theirs in popular music. They can't even be found in school textbooks—our new middle school reading texts feature lots of African Americans and tons of Hispanic Americans, but the only Native Americans in its pages either lurk in the woods to menace poor Christopher Columbus (who, as we all know, "discovered" America) or dutifully aid European settlers. Despite approximately four hundred years of contact here in the eastern United States, the reservation remains a place apart, separated from the dominant culture by the remnants of historical distrust that goes both ways.

Our state has made some inroads in trying to remedy this situation. State Law LD 291 requires that local "Native American history and culture be taught in all elementary and secondary schools." It also requires the state's Department of Education to include local "Native American history and culture in the system of learning results" (thus making Native American history a requirement of assessment under No Child Left Behind). The bill also establishes "a commission to investigate and recommend how the Department of Education will accomplish this task" (http://janus.state. me.us/legis/LawMakerWeb/billtextsearch.asp)

Readers may be familiar with similar legislation in Wisconsin. From the Wisconsin Department of Public Instruction website, Act 31 is "a mandate requiring K–12 teachers to teach about Native American history and treaty rights." This 1989 act was a response in Wisconsin toward developing understanding of treaty rights. It does not, however, mandate that Native American history be included as a requirement of assessment. Other states are making strides toward recognition of and education about local indigenous populations, for example Montana, North Dakota, and South Dakota, but we have not been able to find any other state that includes assessment of learning.

Because of this new state law, the high school closest to the reservation—the one a large percentage of our students attend when they leave our K–8 school—is adding a course on Native culture taught by a member of the tribe. The effects of this course and others like it will take a while to be felt, but we must be optimistic that the end result will be a lessening of the racism our children meet and the estrangement they feel when they venture off the reservation.

So, with a hopeful eye to the future, I take readers back to the present reality of teaching in our school. Perhaps the most difficult challenge we face is the student who does not engage. We can learn our content areas backward and forward. We can model until the cows come home. We can pour our hearts and souls into lessons we hope will captivate and motivate. But there are some students who will refuse for some reason to allow themselves to be caught in our well-crafted webs.

I see these students in my classroom every day. For the most part, they aren't the "behavior" kids. They don't qualify for special services. They are otherwise capable, personable children who often score at or above grade level in math and reading as measured by the computer-accessed testing done in the school in the fall and again in the spring.

What is happening here? Are we teachers failing these students or is American education in general failing them?

Last year, I was a reading and language arts teacher. Fifth and sixth grade reading programs exist, but in my seventh-grade reading class I was on my own. I developed a book list consisting almost exclusively of books written by Native authors, for Native students, and available exclusively from a small company on the West Coast. I never would have known about this resource if I hadn't complained one day to a friend from the university's Native American Studies Department that I couldn't engage my reading students. But now, after achieving Highly Qualified Teacher status for middle school math, I am tasked with seventh and eighth grade mathematics. I can't give that same seventh grade a mathematics text that is relevant to their culture. Nor can I find a good American history text that honestly reflects the history of Native people in this country. Why should it matter? Math is math. History is fact. But maybe not. Maybe math isn't math if your culture has an entirely different way of looking at quantity and measurement. Maybe history isn't fact if the bright lights of your people are overlooked and denigrated.

Maybe some of our most intelligent students are smart enough to see the future ahead of them if nothing changes for indigenous people. Maybe they're refusing to play the education game because they think it will turn them away from their culture. Will I still be Native, they might be asking themselves, if I learn this stuff from off the rez?

I don't know the answers. That's why this project could be so important to us. We need to talk to the students, listen to what they say, and use what we learn to change our pedagogy so that it effectively engages all students. Otherwise we could continue to lose some of our most promising children, and no Native group can keep bleeding children and survive.

STUDENT RESISTANCE TO LEARNING

There is an abundance of literature on student resistance to education (Ganzel, 1998; Garber, 2000; Saunders & Saunders, 2002; Alpert, 1991; Coladarci, 1983; Kitchen, 2003; Raffini, 1986; Moore, 1997; Dehyle, 1992). The realm of training realizes challenges in their environment as do K–12 educators; "one resistant learner can ruin your day. Two or three can make you wonder if you chose the right career" (Ganzel, 1998). We are not challenging the assertion that"…Teachers are not as prepared to deal with students who resist learning yet seem to have the ability to do well in school and apparently choose not to complete assignments or participate in class activities, consequently choosing to earn failing grade" (Garber, 2002, p. 1), but read this with an eye toward an alternate view: maybe some students are victims of structural issues of which they have no control. We agree with D'Ambrosio (1997) "The important point is to create a learning environment in the classroom in which the teacher recognizes that the student has preexisting knowledge, mainly knowledge based on cultural practices. The classroom is a place for the teacher also to acquire knowledge" (p. 246).

Consider a faculty discussion focused on mathematics education during the teaching of the two mathematics courses. The conversation was focusing around accommodations in assessments for children which then quickly turned to the children that "could care less" about their mathematics learning.

> **R¹:** There are um, what's the word I'm looking for, there are adjustments that can be made for students who are labeled as ESL or as special ed or whatever that you can read the math problems to them. I mean to me these are reasonable things for an eighth grader to know. The problem comes with students who have no interest in learning and no interest in becoming educated to an eighth grade level and then that's where the pressure gets turned on me. As a math teacher I have a bunch of students who could care less about percents or fractions or any of this. Although I do think this is a reasonable expectation for them with particular students I don't feel like this is a reasonable expectation for me as a teacher.
>
> **R:** Well if I'm understanding you correctly you're saying the same thing we all worry about, the fact that no matter how well you teach if someone has no interest in learning what you teach they're probably not going to learn it at a level you want them to. The old lead a horse to water bit.
>
> **R:** However there are children who are bright who choose to be bored. They choose not to engage.
>
> **R:** Even if that is something that is very interesting and they normally might find it interesting, they still can choose to stand back and be cool and not engage with the activity (Field notes, fall 2005).

From a multicultural education perspective, which we use in considering mathematics education of our students, D'Ambrosio (1997) reminds us that

> Multicultural education can be successful and give more than lip service to equity and diversity if it recognizes that the practices and perceptions of learners are the substratum on which new knowledge is built. Thus it has to be constructed on the individual and cultural history of the learner and has to recognize the diversity of extant cultures that are present in specific communities. (pp. 245–246)

We hypothesize that a contributor to caring less about learning mathematics has to do with the fact that new knowledge is not built on the existing knowledge of these Indigenous youth. Certainly textbooks are not responsive to local contexts, which we interpret as an educational barrier.

Dehyle (1992) reminds us to consider what barriers our education institutes create, from her work with dropouts, which is frequently "defined as an issue of individual failure" (p. 24). Moore (1997) discussing her undergraduate students: "Resistant students refuse to engage in the content of the course. The consequences, in terms of what students get out of the course and distraction they represent for other students as they attempt to garner support for their resistance, can be substantial" (p. 128). Raffini (1986) speaking to evaluation and norm-referenced examinations offers: "Ironically, one of the problems may be that many students are *not* willing to accept mediocrity, choosing instead apathy and even failure rather than 'average' or 'below-average' performance" (p. 53). Raffini (1986) continues: "When students see school as a threat to their self-worth, some are forced to cho[o]se apathy and noninvolvement as a defense" (p. 53).

Maybe Alpert (1991) was correct: resistance is a product of "teaching approach" (p. 351). Alpert suggests "student resistance is likely to appear in classrooms where academic subject-matter knowledge is emphasized by the teacher and a recitation style is typical of classroom language interactions" (p. 351). We know, as D'Ambrosio (1990) so poignantly stated, that "Science and mathematics education means action. Simply accumulated knowledge in science and mathematics, which easily falls into rote learning, comes closer to history than to true science and mathematics" (p. 375).

Coladarci (1983) reporting his findings, "One should interpret these data cautiously, (p. 17)" of Native American dropouts found the following that we see as pertinent to the resistance we are trying to understand. (Many of Coladarci's 1983 findings were replicated by Dehyle's research in 1992.) The sense that teachers do not care about students found by Coladarci (1983) was also reported by Dehyle (1992), "Navajo and Ute school leavers felt their teachers did not care about them" (p. 28). More encouragement from teachers was cited as a need by Coladarci (1983), and Dehyle (1992) quoted one of her subjects:

> I didn't care to finish high school. It was not that important. You see, I was just learning the same thing over and over. Like the teachers didn't expect anything of you because you were an Indian. They put you in general education, basic classes, and vocation. They didn't encourage college bound classes. (p. 33)

One more consideration has to do with school content. Coladarci (1983) had a data point that "might represent the perception that the curricula did not adequately embrace Native American culture" (p. 20). We are aware of this fault in the curriculum and embracing new ways to be responsive through ethnomathematics: "As a pedagogical programme, ethnomathematics stems from love, respect and solidarity for children and for adults: respect for each person's differences and solidarity with their needs" (D'Ambrosio, 1990, p. 369).

The school's two culture teachers are rich sources of information of how "we" the collective teaching community can and should be making curricular differences for the students. As Raffine (1986) states: "…if our goal is maximum effort from *all* students, then our educational system must demonstrate to all students that increased effort *can* result in success (p. 55)" and we expect the increased effort on the part of faculty, staff, and administration will result in student success.

TOWARDS A SOLUTION

We are ready to accept the challenge offered up by Garber (2002) when she stated:

> Teachers often do not want to hear what resistant learners think about teaching and learning, maybe because they are fearful of what they might hear. As Cuban (1989) said, "The two most popular explanations for low academic achievement locate the problem in the children themselves ('they lack ability, character, or motivation') or their families ('they are poor, lack education, and don't teach their children what is proper and improper in the dominant culture') instead of considering the role of school culture or the structures of the school. (p. 781) (p. 4)

We are striving toward letting "students' motivation to do mathematics grow out of the natural cultural environment. Mathematical explorations should be generated by discussions among the students" (D'Ambrosio, 1997, p. 247). We want to hear what our resistant learners think about teaching and learning. It is important for us to hear how our students perceive the school structure and culture and what we as responsive educators can and must do to engage all our students. "The clear edge that the teacher most often has over the student should be adapted into a congenial partnership, building up into positive self-esteem for the student, and should never reflect an arrogant , imposing, authoritative attitude, which does no more than reinforce negative self-esteem" (D'Ambrosio, 1990, p. 375).To that end we await research approval to begin the interview process.

NOTE

1. Each R is a new voice in a conversation held with faculty and staff.

REFERENCES

Alpert, B. (1991). Students' resistance in the classroom. *Anthropology & Education Quarterly, 22*(4), 350–366.

Coladarci, T. (1983). High-school dropouts among Native Americans. *Journal of American Indian Education, 23*(1), 15–22.

Cuban, L. (1989). The "at-risk" label and the problem of school reform. *Phi Delta Kappan, 70*(10),780–781.

D'Ambrosio, U. (1990). The history of mathematics and ethnomathematics: How a native culture intervenes in the process of learning science. *Impact of Science on Society, 40*(4), 369–377.

D'Ambrosio, U. (1997). Diversity, equity, and peace: From dream to reality. In J. Trentacosta & M. J. Kenney (Eds.), *Multicultural and gender equity in the mathematics classroom: The gift of diversity* (1997 Yearbook ed., pp. 243–248). Reston: National Council of Teachers of Mathematics.

Dehyle, D. (1992). Constructing failure and maintaining cultural identity: Navajo and Ute school leavers. *Journal of American Indian Education, 31*(2), 24–47.

Ganzel, R. (1998). Go ahead, make me learn. *Training, 35*(8), 42–48.

Garber, S. H. (2002, April). *"Hearing their voices": Perceptions of high-school students who evidence resistance to schooling.* Paper presented at the Annual Meeting of the American Educational Research Association, New Orleans, LA.

Kitchen, R. (2003). Getting real about mathematics education reform in high-poverty communities. *For the Learning of Mathematics, 23*(3), 16–22.

Lincoln, Y. S., & Guba, E. G. (1985). *Naturalistic inquiry.* Newbury Park: Sage.

Moore, M. (1997). Student resistance to course content: Reactions to the gender of the messenger. *Teaching Sociology, 25*(April), 128–133.

Raffini, J. P. (1986). Student apathy: A motivational dilemma. *Educational Leadership, 44*(1), 53–55.

Saunders, J. A., & Saunders, E. J. (2001). Alternative school students' perceptions of past [traditional] and current [alternative] school environments. *High School Journal, 85*(2).

Smith, L. T. (1999). *Decolonizing methodologies: Research and Indigenous peoples.* Dunedin: University of Otago Press.

CHAPTER 9

ISSUES OF STATUS AND VALUES IN THE PROFESSIONAL DEVELOPMENT OF MATHEMATICS TEACHERS

Libby Knott
The University of Montana, USA

ABSTRACT

This chapter identifies issues of status that often arise in the classroom. These issues are difficult to deal with and teachers often lack the necessary tools. A professional development (PD) program for K–12 math teachers attempted to address these issues and train teachers how to recognize and to deal with them, by teaching them about community agreements, group roles and protocols for use in small groups. After two years of a PD program, teacher participants reported seeing positive results among their students. The quiet students were speaking up, while the domineering students were learning to allow equal time for all students to have a voice.

International Perspectives on Social Justice in Mathematics Education, pages 139–152
Copyright © 2008 by Information Age Publishing
139

INTRODUCTION

Low status students are often easy to spot in the classroom. They speak less often than their peers, and when they do try to contribute they are more likely to be ignored. Their ideas are seldom heard, acknowledged or valued by their classmates. They are often physically separated from their group and denied participation in the group task, shut out verbally and by the exclusionary body language and other non-verbal cues of their classmates. In reaction to the pain of exclusion they may demonstrate disruptive behavior in the classroom. When teachers recognize the existence of low status students in their classrooms, they are often ill-equipped to deal with them, and unprepared to integrate them successfully into group activities.

In this chapter I address the possible causes of these status issues and suggest several ways to go about resolving the resultant problems. To begin, I will make the connection between status and social justice, and then identify the roots of status in the classroom. I will describe how status can interfere with and even prevent effective group interactions. I will next describe a sequence of effective strategies that were employed in a professional development program with teachers. Lastly I will discuss the results and feedback from the teacher participants.

STATUS AND SOCIAL JUSTICE

It is not possible nor even desirable to eliminate status completely, but it is possible and desirable to nullify its negative effects. Status is an inevitable component of almost any collection of people. It is not an inherently bad thing, for sometimes it provides motivation and promotes the desire to achieve. However, the often negative effects of status lead to social justice infractions.

Cultural norms and the underlying social values associated with race (white is better), ethnicity (WASP is better), socio-economic status (SES), (wealthy is better), gender (male is better,) naturally give rise to status issues in the classroom. Historically, in the US at least, it was often expected that white males would dominate the class time in discussions, achieving more, contributing more, saying more, and doing more than their non-white, non-male counterparts. Such patterns give rise to considerations of social justice, since the accompanying expectations have an insidious way of being met. Teachers tend to call on white males for their contributions more often than they call on blacks or females (Rosenholtz & Cohen, 1985, Cohen, 1994). This is harmful to low status students since consistent findings show that "the student who took the time [or was afforded the opportunity] to explain, step-by-step, how to solve a problem was the student who gained the most from the small group experience" (Cohen, p. 10.) Students who are not afforded the opportunity to explain, or are not confident enough to grasp such an opportunity for themselves, do not gain as much from the same small group experience, and are consequently denied the ability to achieve.

Community agreements or social norms in general, as well as socio-mathematical norms—mathematics classroom rules about expected, desirable behaviors of students when working on mathematics, established by the teacher—establish values in the classroom. For example, in a classroom where the norm is that students will explain their work, then it is clear that in this classroom, explanations—probably both written and verbal—are expected and valued. In other classrooms, it may be the norm that students are expected to submit neat work with correct, boxed answers. In this classroom it is clear that accuracy and neatness in mathematics are valued.[1] But it is possible that these norms and the associated values can be expanded to prevent the ill effects caused by status that operate against all students having a fair chance to excel.

If students are actively prevented from participating in the mathematical activity of the classroom—shut out of investigations, unable to contribute, not able to handle manipulatives, not listened to, it is perhaps being enabled by teachers who are insensitive to the consequences of status issues. This exclusionary behavior is less likely to be observed in a traditional, lecture-style classroom where the teacher does most of the talking and students work individually and quietly on given tasks. In this environment students' work and individual contributions are less valued. That does not mean that such exclusion is not taking place—it is simply not so obvious or immediate, and much less likely to be corrected.

In contrast, in the ideal mathematics classroom environment where open-ended mathematics tasks, large-scale inquiry and modeling activities are stimulated, and where participation and communication via small-group activities and through using manipulatives are encouraged, this kind of status-differentiated behavior is more likely to be observed. It can, however, be addressed and mitigated by careful and thoughtful classroom management. The next section addresses some specific causes of status issues that operate during small group work in mathematics classrooms.

THE CAUSES OF STATUS ISSUES
IN SMALL GROUP ACTIVITY

Cohen (1994) has noted in the working of small groups that individuals seldom share the talk time equally. Most often, some group members will talk a lot, while others contribute little or nothing to the discussion or task.

Merely arranging students in groups is no guarantee that meaningful work takes place. In fact Sfard (2000) and Kieran (2001) found that students will often talk past each other without ever engaging in meaningful discourse. In addition, group members themselves associate those who talk more with higher achievement and those who talk less with lower achievement, thus setting up expectations for future participation and achievement of group members and perpetuating the disparity. Cohen (1995) identifies five types of status issues occurring in the classrooms.

The first type is *expert* status, where some individual is recognized as an expert on the particular subject area. This person is not necessarily the brightest, most able student in the classroom, but has for some reason been labeled as expert by his/her peers. This person may dominate the group and will be deferred to for the answers. Students are usually aware of the grades and relative student rankings of all their classmates and will readily defer to the student who is commonly perceived as being the top student in the class. Expert status has traditionally been conferred as an award for achievement and hard work. Expert status exists and we do not want it to go away. The problem that must be addressed is how that "expertness" is defined, acknowledged and rewarded, and by whom. And our very definition of what constitutes "expertness" should be examined.

Cohen (1995) also identifies *reading ability* status. A student with high reading ability is often labeled as a high status individual, and is deferred to in classroom activities and discussions, even when the particular activity does not require special reading ability. Rank status according to reading ability is usually common knowledge in the classroom.

> This means that if you are a poor reader, it is not only you who expect to do poorly—all your classmates expect you to do poorly as well! It is an unenviable status, particularly when one thinks of how many hours a day you are imprisoned in a situation where no one expects you to perform well. (Cohen, p. 30)

There is the erroneous perception that high ability in reading implies high ability in other tasks that have little or nothing to do with reading per se.

A third area, *peer status*, derives from social standing due to attractiveness, popularity, maturity or accomplishment in sports, for example. Students who have high peer status in the classroom are more likely to seek out and dominate small group tasks in non-related areas.

Societal status, based on general cultural beliefs and values, also comes into play in the classroom. In most Western societies it is customary that males, and particularly white males, have higher social status than those who are of minority race and/or female. Studies have verified that "men are more often dominant [in the US] than women in mixed-sex groups and Anglos are more often dominant than Mexican-Americans who have an ethnically distinctive appearance" (Rosenholtz & Cohen, 1985, in Cohen, 1995, p. 32).

Socio-economic ranking outside the classroom also comes into play in the classroom, with children from poor families being routinely assigned low status, while those from wealthy families are assigned high status. This has been the justification for requiring school uniforms, for instance.

EXPECTATIONS AND STATUS CHARACTERISTICS

Status, whether derived from race, gender, social class, socio-economic standing, reading ability or attractiveness, is not altogether bad. It creates expectations for performance and encourages competence for some. High status students are expected to begin contributing immediately to the group process—they are social leaders. Not only are these students expected and encouraged by their peers to achieve, but they quickly come to expect this of themselves and thus begin a display of competency immediately. This quickly builds on itself. In contrast, low status students are not expected to make significant contributions in group-work nor do they expect it of themselves. Typically they are quiet and contribute little or nothing to the task. This cycle is self-perpetuating, and unless the teacher defeats these negative consequences, serious social justice issues may arise.

There are two primary factors that influence the workings of a small group: the nature of the group task itself and who participates frequently at the outset of the group activity. This is true in school and college classrooms even among students with records of high achievement. Status issues interfere with voluntary participation. It has been well-established that those individuals who "start talking right away, regardless of their status, are likely to become influential" (Cohen, p. 35). It is clear that status issues create a self-perpetuating spiral of positive effects for high status individuals, and of negative effects in the case of low status individuals. A case of "the rich get richer". To break the spiral, teachers must learn to deal with these issues.

DEALING WITH STATUS ISSUES AS A COMPONENT OF A PROFESSIONAL DEVELOPMENT PROGRAM

Students do not intuitively recognize or display the skills they need to be successful participants in and contributors to group activities. They must be taught these skills by their teacher, who may not be conversant with successful strategies. To address this need amongst teachers, awareness in recognizing and dealing with status issues in small group activities was integrated into a large, federally and state funded professional development program for mathematics and leadership development for K–12 teachers in a western state. I will discuss each of the implementation phases, and follow up with a discussion of changes that occurred among the participants.

Pre-Institute Preparation of Faculty

The faculty instructors in the professional development program (this author was one) who were to provide the mathematics content and the leadership classes for the teacher participants, were a diverse group of individuals selected from district leadership roles, community college, four year public and private college and university faculty from both the mathematics and education communities. Before it was possible for them to work with a large group of K–12 teacher participants, they themselves had to become sensitized to status issues (including their own) and learn how to deal with them.

The first task for the faculty as a group was to arrive at a set of community agreements that would provide the basis for all future interactions. Once prepared, these agreements were prominently displayed at each subsequent meeting, and reviewed frequently. This was a crucial piece of the fabric that would provide a safe environment for people of differently perceived status to participate confidently in the program. Once these were established among the instructors, they were used as the foundation for building a set of community agreements for the teacher participants.

The faculty instructors worked from the premise that, during our time together, we needed a safe and open climate where we would be able to reflect deeply about important ideas related to mathematics, teaching learning and leadership; to work together on complex questions and issues that may challenge our beliefs and practices; to share points of view; and to examine our practices. In this climate we began working towards establishing the community agreements to which we would adhere throughout the program. The main agreements that evolved from our discussion were:

- Be willing to focus, reflect, listen and share.
- Encourage others with positive feedback; suspend negative judgment.
- Be respectful of differences (and wrong answers); ask permission to suggest corrections.
- Acknowledge new ideas and contributions (without interruptions).
- Give your ideas voice without fear of judgment or reprisal.
- Present ideas in a non-threatening (non-demeaning) way.
- Allow and encourage others to speak and ask questions.
- Challenge your own opinions and thinking.

- Take time to process (without side-bar comments).
- Come to class prepared to work and learn.
- Keep in mind your long-term goal to achieve.

This meant that in our professional learning community we agreed to be respectful of each others' ideas, questions and thinking, we would honor and appreciate diversity, embrace change, listen to understand, take risks, collaborate, be introspective, be accountable and have fun.

This heterogeneous group of professionals then engaged in workshops about best practices—how to model what was expected of the teacher participants themselves, and what those "best practices" would look like. Since a crucial component of the best practices model we adopted involved small group work dealing with investigative activities, we focused on strategies for managing small groups. Material from *Designing Groupwork: Strategies for the Heterogeneous Classroom* (Cohen, 1994) was discussed and modeled. The accompanying video *Status Treatments for the Classroom* (Cohen, 1994) was used to provide vivid examples of problems related to status, and as a point to initiate our discussions. We developed the skills we would need for building more effective group participation in the classroom that would mitigate inherent status issues among the K–12 teacher participants during the summer Institute. This was not an easy task.

Learning about Effective Small Group Work

Establishing community agreements and discussing classroom norms explicitly provides an environment where low status students can have the chance to participate, their voices can be heard, and they can feel valued as important contributors to the tasks. But this would not by itself change the low expectations that others have of the competence of low status students, nor would it change their self-perception of their own abilities. We had to establish expectations for competence and get buy-in from the teacher participants. We learned to do this by paying very careful attention to the design of the mathematical tasks we would use; ones that had multiple entry points and could be worked on meaningfully by teacher participants of varying abilities. We designed activities that required conceptual understanding rather than higher math prerequisites and did not rely heavily or exclusively on algebraic manipulations. This was important since we suspected that many of the elementary teachers would not feel confident of their algebraic skills, whereas secondary teachers would use this approach freely, and often as a first choice. Since all teachers could enter the problem in a meaningful way, no-one would feel excluded.

This preparatory work included an initial phase of identifying and describing status issues, explicitly discussing them, and instructing the faculty in ways to recognize and address them. Only after this detailed step-by-step practice by which instructors in the program became proficient at understanding and modeling ways to deal with status issues in faculty meetings and in the classroom, were they able to then model the desired behavior in the classes with the teacher participants.

Skill-Building Exercises for Small Group Effectiveness

The key to successful small group activities that diffuse status issues lies in giving everyone a role to play. We learned that it was necessary to assign group roles to teacher participants every day, and rotate both the groups and the roles that individuals played in those groups at least daily and sometimes more than once during each 2½ hour class period. By establishing cooperative norms such as "everyone participates, everyone helps" and assigning group roles that ensured everyone had something to do, we would diminish the status issues that might otherwise have been insurmountable.

We rehearsed the guidelines for assigning students to groups, and for assigning roles to each group member. We established four roles that we would use for teacher participant group members: facilitator, recorder/reporter, resource monitor, and team captain. We reviewed the responsibilities of each job: facilitator was responsible for getting the team conversation started, and for making sure that everyone understood the task. The recorder/reporter was responsible for giving update statements on the team's progress, for organizing and introducing the report on the group's activities, and for making sure that everyone on the team was recording necessary information in their journal. The resource monitor was responsible for collecting material and resources that the team needed for the activity, calling the teacher if there was a team question, and organizing cleanup. The team captain was responsible for enforcing use of norms and encouraging participation, for finding compromises, and for acting as a substitute for absent jobs (there would sometimes be only three members in a team). See Appendix II for a complete description of these team roles.

This however, is not sufficient to ensure the smooth workings of small groups without any of the debilitating influence of status issues. Simply assigning participants to groups is no guarantee that they will work in the expected way. The Institute faculty had to also learn about and implement group protocols to ensure that everyone in the small groups had the opportunity to participate. We practiced and prepared to implement group protocols such as 'private think time', 'dyad sharing', 'go around protocol', 'popcorn share', 'jigsaw share', and 'whole group share'. To become conversant in these group strategies we practiced them amongst ourselves at every planning meeting prior to the Institute, but it wasn't until we actually implemented them that we built a really solid foundation. When I recall how much effort it took for these professionals to be effective, I realize how much of a challenge it is for any K–12 teacher to be successful at managing group work.

TEACHER PARTICIPANTS

During the Institute, with K–12 teacher participants from diverse geographical regions, from both urban and rural settings, in the same classroom for the content and leadership classes, issues of status immediately arose. Fortunately, because of their previous training, the faculty knew how to respond. We addressed these issues by cooperatively establishing a set of community agreements that both faculty and teacher participants would adhere to, and a set of agreements (social and socio-mathematical norms) for the mathematics classroom environment. The community agreements that we established were very similar to the ones that the faculty had established for use amongst themselves. We agreed that everyone had the right to be heard; that we all had the duty to listen for understanding (as opposed to listening to respond); that we would not refer to the grade level that we taught; that we would not tolerate any negative self-talk (e.g., I can't do this). The full text of our community agreements may be found in Appendix 1.

The social and socio-mathematical norms we established were that everyone had the right to ask any question; the right to an answer to their mathematical question; that mathematical answers, both correct and incorrect or incomplete, were accepted gratefully; that incomplete or incorrect answers would become a step to build on to further knowledge; that we would try to justify all mathematical comments; that we would try to generalize our mathematical findings. We kept our agreements visible at the front of the room and reviewed them on a daily basis throughout the three-week Institute.

First Year Results

In this PD program, the program coordinators had borrowed from the 'best practices' camp—namely, by requiring faculty to model the practices that they wanted to see the teacher participants use in their own classrooms. Faculty necessarily addressed status issues directly and explicitly as part of the program. Because the program involved teachers from kindergarten through twelfth grade taking content mathematics classes together, and also involved these same teachers participating in a leadership class with each other and their own administrators, status issues were bound to occur.

During the first year of the Institute we did not lay a solid foundation and as a result we heard some grumblings from some teacher participants who felt they were being dismissed by their peers as not knowing mathematics since they were "only" elementary school teachers. The grumblings were almost entirely absent during the second year.

There were tears and frustrations, there was joy and fear, to be sure. But with the agreements established and clearly in place in the classroom, and reviewed *every day*, the teacher participants learned how to respond reflectively, and were able to step out of their habitual behaviors and recover quickly to minimize the negative effects of status.

During the summer Institute, the faculty systematically enforced group roles and responsibilities in the content classes. At first we assigned the teacher participants randomly to groups and within each group established roles for each group member. Sometimes this enforcement was over the objections of the teacher participants: "Why do we have to keep doing these jobs? This is silly!" they would complain. "Why must we keep reviewing the responsibilities of each group member?" Eventually, it turned that they would thank us for our persistence in enforcing group roles. They came to see the importance of team roles and responsibilities in "evening the playing field", providing an environment in which all students were given a voice, all students were able to participate, and all students were valued. They were also grateful for having practiced them so much; many said it made it so much easier to implement in their own classrooms, since it had become automatic for them. There were times when the group roles broke down, and then the teacher participants would experience the discomfort of not being able to contribute, of not being listened to, of being shut out of the activity. This had a lasting impact on them. They were determined that their students would not have these negative experiences, and were convinced that effective group management was the key.

Second Year Results

In year two of the project, many participants reported feeling far less intimidated and/or insecure about their participation with other K–12 teachers. This was due, in large part, to the excellent preparation in status issues that the teacher participants and the faculty had received. Our work in modeling and enforcing small group or team roles was critical. All of the classroom activities were designed with small groups in mind. We used random groups of size four as much as we could, and enforced and rotated group roles of team captain, team facilitator, team recorder/ reporter, and team resource monitor. During this second year, the teachers complained far less about our strict enforcement of the group roles. In their evaluations, teacher participants reported that their experiences put them in touch with how their students must feel in similar situations, when someone of relatively high status prevails over someone with lower status, when they felt ill-equipped to answer the questions, and were somewhat intimidated by the higher status teachers, often teachers of higher grades. All felt that the challenge was a big one, but everyone reported learning gains from the experience. They also shared stories about the changes in their classrooms because of their awareness of status issues, and how to deal with them.

We found that during the second summer, the consequences of status issues were greatly reduced from the previous year. When we did group work, the teacher participants were respectful of everyone's thinking. One teacher responded that during the first summer, she had found it difficult to work with many of the high school/middle school teachers as she often didn't have time to adequately process information before they had the answer. She said that she found she had to move on before she had a good handle on the concept. The second summer was different for her. 'Private think time' afforded her the opportunity to consider the problem before discussing it with her group. Teachers for whom English was a second language reported that they appreciated the group protocols that were established during the Institute. One of their concerns had been that they would not have sufficient time to understand the task before being expected to complete it. With protocols such as 'private think time' firmly in place, the small group facilitator understood that it was his/her job to make sure that private think time was enforced prior to engaging in each group activity, so that every participant had time to read the directions and secure a clear understanding of the task.

Back in their classrooms, the teacher participants reported that it was important that they had been required to practice these roles, internalize them, and have sufficient familiarity with them so as to make it natural for them to implement them smoothly in their own classrooms. Enforcing these group roles was the single most reported, and criticized, but helpful strategy learned during the Institute that would help them address status issues in their own classrooms later on. One teacher reported "I try to create a climate that celebrates wrong answers as a place to start new learning. I also need to implement more protocols into my lesson plans and make sure that I enforce the classroom agreements." Clearly for this teacher the classroom agreements and the group protocols have had a lasting impact on her every day teaching.

CONCLUSIONS

Status issues are present in almost every setting where people get together in groups. It is important to realize that these issues are ubiquitous, and that we need to be shown how to nullify their effects. In our work with K–12 teachers, through modeling and practicing behaviors that work we have made significant progress towards addressing this issue. Teachers report increased discourse in their mathematics classes as all students participate more fully and freely in group activities. Imagine a world in which small groups of people working together understand the issues of status, and know how to address them, where their time is spent devising explicit community agreements and reviewing group roles and agreements frequently during their work, and where group protocols are used freely as they carry out their assigned tasks.

NOTE

1. For more research on the study of values in the mathematics classroom, see Bishop, 2000, (TCM article date), 2002, Bishop, Fitzsimmons, Seah, Clarkson (1999), Clarkson, Bishop, Seah, FitzSimmons 2000.

APPENDIX I

Community Agreements—Expanded Version

We agreed to be **respectful of each others' ideas, questions and thinking** by recognizing that everyone had something to contribute. Each participant agreed to be an *active* participant; we agreed to give everyone a chance to lead; we agreed to interact with interest; to be encouraging; to respectfully discuss disagreements about ideas; to use constructive problem solving and feedback; to allow time for reflection and think time by all; to honor our own ideas and thoughts; to assure all the freedom to work without fear of reprisal; and to present ideas in a non-threatening manner.

We also agreed to **honor and appreciate diversity** by providing an environment where all voices can be heard and listened to; to allow for and support everyone, regardless of their mathematical backgrounds; to accept that we are all at different levels in our mathematical knowledge; to embrace different opinions and perspectives and learn from them; to value all ideas and suspend judgment; and to strive for equitable participation.

Further, we agreed to **embrace change** through being open-minded; being open to change; being open to new ideas about learning, teaching, and mathematics; challenging our thinking; maintaining a positive attitude towards self and others and to the reasons we are here together; and by taking time to process events.

We agreed that listening was key, and so we agreed to **listen to understand** by being an active listener; by asking clarifying questions; by allowing time for reflection; by allowing everyone to have a voice without judgment; by allowing for thinking and reflection time; by asking genuine and thoughtful questions; by limiting side-bar conversations; and by listening to others without interruption.

We agreed that we would **take risks**, be willing to cross grade level lines to learn from others; that risk taking was okay by remembering that failure and discomfort are part of the learning process; that we would validate other points of view; share concerns openly; and respond supportively to other ideas.

We agreed to **collaborate** with each other by working together to increase understanding and by acknowledging that this was the responsibility of all group members; to work cooperatively; to allow everyone to participate, but to not force those who weren't ready; to remember that all individuals need to work in the group; to respect everyone's need for think time and processing; and to celebrate our own and each others' AHA!s.

We agreed to **be introspective** by being non-defensive and reflective about our own practices; by being responsible for our own learning, thoughts, actions and participation; by being present and positive; by being willing to be wrong and to compromise; and by examining discomfort.

We agreed to **be accountable** by honoring time commitments; by being responsible participants; by being present and being prepared; by making what we learned useful to take back to students and colleagues; by participating and sharing knowledge, ideas, learning strategies, and learning styles with each other.

Last of all, we agreed to **have fun** by having a sense of humor and laughing with each other!

APPENDIX II

Team Jobs and Responsibilities

Facilitator:
- Gets the conversation started
- Makes sure everyone understands the task

Sample questions: Does everyone get what we are supposed to do? Are we ready to go on to the next part?

Recorder/Reporter:
- Gives update statements on team's progress
- Organizes and introduces the report
- Makes sure the team is recording in their journal

Sample questions: We need to keep moving so we can.... Did everyone get that information? Slow down, I need to get this recorded.

Resource Monitor:
- Collects material and resources that the team needs
- Calls the teacher if there is a team question
- Organizes clean up

Sample questions: Do we all have the same question? We need to clean up. Can you... while I...?

Team Captain:
- Enforces the use of norms and encourages participation
- Finds compromises
- Substitutes for absent jobs

Sample questions: We need to work on listening to each other; Remember, no talking outside our team; Let's find a way to work this out.

REFERENCES

Cohen, E. (1994). *Designing groupwork.* New York: Teachers College Press.

Cohen, E. (1994). *Status Treatments for the Classroom* Video available from Teachers College Press, New York.

Kieran, C. (2001). The mathematical discourse of 13-year-old partnered problem solving and its relation to the mathematics that emerges. *Educational Studies in Mathematics* 46(1–3), 187–228.

Rosenholtz, S. J., & Cohen, E. G. (1985). Activating ethnic status. In J. Berger & M. Zelditch, Jr. (Eds.), *Status, rewards, and influence* (pp. 430–444). San Francisco: Jossey Bass.

Sfard, A. (2000). Steering (dis)course between metaphors and rigor: Using focal analysis to investigate an emergence of mathematical objects, *Journal for Research in Mathematics Education* 31(3), 296–327.

CHAPTER 10

CONNECTING COMMUNITY, CRITICAL, AND CLASSICAL KNOWLEDGE IN TEACHING MATHEMATICS FOR SOCIAL JUSTICE

Eric Gutstein
University of Illinois—Chicago, USA

ABSTRACT

In this article, I describe conceptually, and give an example of, an aspect of teaching mathematics for social justice—teachers' attempts to connect three forms of knowledge: community, critical, and classical. The setting is a Chicago public high school, oriented toward social justice, whose students are all low-income African Americans and Latinas/os. Drawing from the experience of creating and teaching a mathematics project that emerged from a central disruption in the life of the school community, I discuss complexities and challenges of creating curriculum from students' lived experiences that simultaneously develops their critical sociopolitical consciousness and mathematical proficiencies.

International Perspectives on Social Justice in Mathematics Education, pages 153–167
Copyright © 2008 by Information Age Publishing
153

INTRODUCTION

Teaching and learning mathematics for social justice has its roots in the mathematics education work of Skovsmose (1994, 2004) and Frankenstein (1987, 1998), among others. It builds on work in critical pedagogy, in particular, Freire's (1970/1998) and others such as Giroux (1983) and McLaren (2007), and also draws upon culturally relevant pedagogy (Ladson-Billings, 1994, 1995b; Tate, 1995). Though proponents and researchers describe it in different ways (e.g., some refer to it as "critical mathematics"), there are certain common pedagogical aims. Two of the most central are that students develop both critical consciousness and mathematical competencies, and there is also the view that these two areas of learning need to be dialectically interwoven by both teachers and students in a conscious manner. That is, mathematics should be a vehicle for students to deepen their grasp of the sociopolitical contexts of their lives, and through the process of studying their realities—using mathematics—they should strengthen their conceptual understanding and procedural proficiencies in mathematics. One of the principal ways for teachers to support students in moving toward these interconnected goals is for the students to engage in mathematical investigations in the classroom of specific aspects of their social and physical world (see Gutstein & Peterson, 2005 for reports by K–12 teachers on efforts to do so).

There are few extended studies of teaching and learning mathematics for social justice in K–12 urban classrooms (Brantlinger, 2006; Gutstein, 2006c; Turner, 2003). These reports shed light on the complexities of enacting critical mathematics pedagogy and certainly point out some of the difficulties in what is mostly uncharted territory. In this brief article, I will highlight one particularly challenging quandary and illustrate it with a short vignette. There is much work to do in theorizing and practicing social justice mathematics, and my purpose here is to point out some issues that I believe currently face those of us who want students to learn mathematics as a vehicle for social change. The matter I discuss is the complexity of building on students' and communities' knowledge while simultaneously supporting the development of their mathematical competencies and critical awareness. I examine it from the perspective of my own work in Chicago (and its public schools) where I have lived, worked, and taught for the past 12½ years, first teaching my own middle-school mathematics class for several years, and for the past few years, working with a new social justice high school in mathematics classes.

CONNECTING COMMUNITY, CRITICAL, AND CLASSICAL KNOWLEDGE

We[1] have adopted a framework in the school's mathematics team of trying to synthesize what we call *community, critical,* and *classical knowledge* (Gutstein, 2006c), or the "three C's." These concepts are not new, but their interrelations have been under-elaborated with respect to mathematics education. We recognize that these may be contested definitions, and we consider the categories (and our thinking) to be provisional and fluid. By *community knowledge,* we mean several different but related components of knowledge and culture. It refers to what people already know and bring to school with them. This includes the knowledge that resides in individuals and in communities that usually has been learned out of school (e.g., their *funds of knowledge,* Moll, Amanti, & González, 2005). It involves how people understand their lives, their communities, power relationships, and their society. We also mean the cultural knowledge people have, including their languages and the ways in which they make sense of their experiences. Some refer to this as "indigenous knowledge," "traditional knowledge," "popular knowledge," or "informal knowledge" (including with respect to mathematics, e.g., Knijnik, 1997; Mack, 1990). Two examples serve to illustrate our meaning. In *Rethinking Columbus,* Tajitsu Nash and Ireland (1998) describe the knowledge of a typical Amazonian elder, who

> …has memorized hundreds of sacred songs and stories; plays several musical instruments; and knows the habit and habitat of hundreds of forest animals, birds, and insects, as well as the medicinal uses of local plants. He can guide his sons in building a two-story tall house using only axes, machetes, and materials from the forest. He is an expert agronomist. He speaks several languages fluently; knows precisely how he is related to several hundred of his closest kin; and has acquired sufficient wisdom to share his home peacefully with in-laws, cousins, children, and grandchildren. Female elders are comparably learned and accomplished. (p. 112)

The other example is one from Freire's *Pedagogy of Hope* (1994, pp. 44–49). In it, Freire recounted a conversation with a group of Chilean farmers. They were having a rousing discussion when the farmers suddenly silenced themselves and asked the "professor" (i.e., Freire) to tell them what he knew. Freire wrote that he was unsurprised by this, having experienced it before, and proceeded to challenge the farmers to a game. They were to stump each other with questions that the other could not answer. Freire went first and asked, "What is the Socratic maieutic?" The farmers laughed, could not answer, then baffled Freire with the question, "What's a contour curve?" The game continued, each stumping the other, until finally the score was 10–10. The point was clear—Freire's knowledge and the farmers' knowledge were both valid and valuable. Each knew things that the other did not; each had to respect the others'—and their own—knowledge. What the farmers knew, from years of shared lived experience, is what we term community knowledge.

Critical knowledge is knowledge about the sociopolitical conditions of one's immediate and broader existence. It includes knowledge about why things are the ways that they are and about the historical, economical, political, and cultural roots of various social phenomena. Various authors (e.g., Giroux, 1983; Macedo, 1994) described *critical literacies*, and we essentially mean the same idea. Freire (Freire & Macedo, 1987) referred to it as "reading the world." In his earlier work on literacy campaigns, he discussed *culture circles* in which groups of workers, peasants, and farmers studied *codifications* (representations of daily life, usually pictorial) and reflected on their meanings (Freire, 1970, 1973). Those sessions allowed the culture circle members to examine their lives from different perspectives, and the process of collectively decoding the representations led the individuals to deepen their understanding of the phenomena. Freire's pedagogy thus provided the opportunities for people to transform their community knowledge about the everyday world that they had often normalized (e.g., we have no work because there are no jobs) into critical knowledge about the same situations (e.g., we have no work because those in power control the distribution of jobs and land, and it is to their advantage to keep some of us unemployed).

It is often the case that community knowledge already is critical, but context matters. For example, relatively young adolescents (e.g., middle-school students) may have knowledge about their life situations, but it is not often critical. Whether or not it is critical depends on several things, including their experiences, those of their families and communities, the level of political consciousness at the time, and the strength of existing social movements. In contrast, adults who are engaged in various struggles may have community knowledge that is quite critical. As an example, a battle is currently taking place in Chicago to stop the displacement of low-income people of color (in particular, African Americans) through gentrification (Lipman & Haines, in press). Many adults in the affected communities have a clear and critical understanding of the political forces allied against them, including their geneses and various forms of subterfuge. I have heard parents in communities where public housing has been demolished (and not replaced) and schools closed (and reopened for "new" residents) eloquently elaborate who and what forces are responsible for their removal, and why. So the lines between community and critical knowledge are not always clear. A major thesis of Freire's work is that *problem-posing* pedagogies can present life situations back to people (whether in or out of school) so that they may pose questions themselves and transform their community knowledge into a more critical state, and consequently be drawn into action to challenge unequal, oppressive relations of power.

The lines between classical and the other forms of knowledge are not so clear either. *Classical knowledge* generally refers to formal, in-school, abstract knowledge. Our focus in terms of classical knowledge is that students have the competencies they need to pass all the gatekeeping tests they will face and to have full opportunities for life, education, and career choices. Classical mathematical knowledge clearly has high-status in society as many have commented (e.g., Apple, 2004) as well as a strong Eurocentric bias (Frankenstein & Powell, 1994; Joseph, 1997). Nonetheless, while we critique it, we recognize its power and cultural capital and argue that students need to develop it for several reasons. They need it for personal, family, and community survival, especially for students who come from economically marginalized spaces. But even more than that, we believe it is crucial that students appropriate, in this case, the "master's tools" with which to dismantle his house (cf. Lorde, 1984). We subscribe to Freire and Macedo's (1987) orientation toward what they referred to as "dominant" knowledge:

> To acquire the selected knowledge contained in the dominant curriculum should be a goal attained by subordinate students in the process of self and group empowerment. They can use the dominant knowledge effectively in their struggle to change the material and historical conditions that have enslaved them. (p. 128)

To connect the three types of knowledge is no simple matter for many reasons. First, there is the question of how might teachers learn students' community knowledge. In Brazil, where Freire and others practice(d) these ideas, the process by which teachers investigate the *generative themes*[2] of a community is complicated. In Porto Alegre's *Citizen School Project*, there is a lengthy and involved ten-step process through which teachers, in collaboration with neighborhood adults, study community knowledge to develop school-wide, interdisciplinary curriculum based on the generative themes (Gandin, 2002). Freire (1970) elaborated his view of how researchers might investigate the themes within a specific community, and this also involved a detailed, multi-step process. There are still more issues, such as the question of how might teachers study community knowledge when they are outsiders to the community, language, and culture of their students (Delpit, 1988), or the fact that the generative themes identified by neighborhood adults may not coincide with those of the youth in schools (I. Martins de Martins, personal communication, July 2003).

Once educators begin to have a grasp of the community knowledge of their students and their families, then they can try to create curriculum, based on those themes, that will support both the development of critical and classical knowledges. This also is quite complicated. First, there are the time constraints imposed on teachers and their working day (which also affects their capacity to investigate generative themes, although in Porto Alegre, teachers were paid for that work). When do teachers have the time to develop new innovative curriculum, let alone cope with all the other demands of teaching? For example, creating standards-based reform mathematics curricula in the U.S. took massive amounts of time, money, and people. The reform curriculum with which I am most familiar, *Mathematics in Context* (MiC) (NCRMSE & FI, 1997–8), required perhaps $8 million, 5 years, and close to 50 people working in two countries before it was fully operational. It is true that MiC was a connected, cohesive curriculum spanning four years (grades 5–8), and obviously developing curriculum for just one school community would require less time. But the time and people power alone needed to create quality curricula testify to the necessary resources required.

Second, to develop curriculum requires a different knowledge base than teaching, despite the interrelationship of the two. My personal knowledge of MiC's development and my professional judgment suggest that there are talented curriculum designers who would have difficulty teaching MiC in urban classrooms because, for example, they may not connect that well with the students nor their communities. This is also probably true for other successful curriculum projects whose authors are primarily university-based mathematics educators. Conversely, there are successful mathematics teachers in urban schools who do not have the knowledge to create rich mathematics curriculum.

Third, successfully navigating the requirements of a standards-based mathematics curriculum is difficult enough, especially under the pressure of neoliberal accountability constraints like the *No Child Left Behind* legislation in the U.S. that mandates repeated testing. But to do so while simultaneously providing opportunities for students to develop critical knowledge in mathematics classes is an added layer of complexity (Brantlinger, 2006; Gutstein, 2006c). It is generally accepted that good (mathematics) teachers need to have content knowledge (Hill & Ball, 2004), pedagogical content knowledge (Shulman, 1986), and knowledge of students and their communities (Ladson-Billings, 1995a, 1995b); but in addition, to develop critical knowledge, teachers also need deep knowledge of social movements, history, culture, political economy, and local and global sociopolitical forces affecting students' lives, as well as particular dispositions toward social change and the politics of knowledge (Gutstein, 2006a). Even when teachers do have these various knowledge bases, ensuring that the mathematics does not get lost when developing critical knowledge and supporting students' sociopolitical consciousness (in mathematics class) is no easy task—the dialectical interrelationships are complicated and more attention needs to be focused in this area, and more experience accumulated (Brantlinger, Buenrostro, Gutstein, & Mukhopadhyay, 2007).

In short, for many reasons, it is quite complex to create curriculum that starts from students' and their communities' lived experiences/knowledge and then simultaneously and with rich interconnections supports *both* mathematical power/ classical mathematical knowledge *and* a critical awareness of one's social context. No such mathematics curriculum currently exists that is broadly applicable partly because of the specificity of local situations, although there are several examples of projects and units of social justice mathematics that have been taught in urban schools (see, for example, Brantlinger, 2006; Frankenstein, 1998; Gutstein, 2006c; Gutstein & Peterson, 2005; Osler, 2006; Turner, 2003). It will not be easy to create high-quality social justice mathematics curricula that teachers can adapt to their local settings, and even allowing for good curricula, the school change and professional development literature is clear that curriculum alone does not ensure effective and appropriate teaching—nor real learning (Fennema & Scott Nelson, 1999). Efforts to work on connecting the "three C's," however we describe them, are needed, and how to do so is an open question with respect to both theory and practice.

An Example of Connecting The Three C's in Practice

I turn now to a short example of our work in a Chicago public high school for social justice in which we attempted to connect community, critical, and classical mathematical knowledge (see Gutstein, 2006b, for details). Briefly, a new school was built and opened in Fall 2005 after a group of residents in a Mexican immigrant community (Chicago's *Little Village*) went on a 19-day hunger strike in 2001 (Russo, 2003). The residents struck for a new school for their community; the school board promised it, then reneged; and the hunger strike was the culmination of a multi-year struggle for a new school in the overcrowded neighborhood. The new school building houses four small schools, each with a maximum of 350–400 students, and each with a different community-determined theme. The school I work with is the social justice high school (known to most as "Sojo").

Although Little Village is overwhelmingly Mexican, the Chicago public school (CPS) board, under a 1980 federal desegregation mandate, racially integrated the open-enrollment, neighborhood school by drawing the attendance lines into a bordering African American community, North Lawndale. Thus the schools are 30% African American and 70% Latina/o. However, by changing the attendance boundaries, the school board also limited Latina/o enrollment, causing friction for some Little Village residents who saw their children's spots in the new building "taken" by African Americans from North Lawndale. Furthermore, given Chicago's history of segregation, racist exclusion, and neighborhood and turf lines, there is an ambivalent relationship between the two communities. Students for the most part intermingle and work together in the school, although there are real tensions outside in the neighborhood.

In January 2006, during the first year when each school had about 100 ninth graders, a local Latino politician held a press conference and proposed a public referendum that the boundaries be redrawn to exclude North Lawndale African American students. Black students, understandably angry, hurt, and scared, immediately went to teachers to voice concerns about being removed from the school. Our mathematics team, on the initiative of one of the math teachers, quickly developed a mathematics project (the "Boundaries Project") whose central question was this: What is a fair solution for both communities?

While our assessment is that there were weaknesses in the project (e.g., we threw it together in two days because of the immediacy of the issue, and it was not clear how much mathematics students learned), our analysis also suggests that there were some considerable strengths. Most notable was that students were quite engaged, and we believe this is because the work students did was genuine. No one knew (or knows) the answer to the central question because, in fact, the solution to the problem has to be eventually determined by the two communities working together in concerted effort to ensure that there are enough spots in quality schools for all the students—something that is not the situation now, even with the new school. The project tied directly into students' lived experiences and generative themes—that is, it built on students' (and their families') community knowledge. The issues of interconnections between the two neighborhoods, their histories, and students' stereotypes toward each other all surfaced. Politically, the two main points with which we wanted students to grapple (i.e., the development of critical knowledge) were that the differences between the communities were far outweighed by the commonalities, despite historical divide-and-conquer techniques used to pit communities of color against each other, and the above point that ultimately there were not enough quality schools for all the students. Mathematically, we asked students several questions about the numbers of Black and Brown students in the building at full enrollment given ratios different than the current 30:70, and the probability of a student from each community being accepted in a lottery (using different possible ratios). We also had them study census tract data and consider how to enlarge the boundaries in North Lawndale so that students from there would have the same chance to be accepted as the Little Village students. This entailed calculating acceptance probabilities for both communities, with various ratios of African American and Latina/o students—and this was further complicated mathematically because each neighborhood has different numbers of high-school aged students. Students also examined data for other nearby schools, as well as local area maps, and overall, they mathematized the central problem of having one new school building for too many students from two different communities. In our assessment, the complexity of the mathematics lay more in this requirement to draw out the mathematical components of the situation, than in any specific subpart or individual problem within the project.

While we know that a week-long project can have only limited impact, we locate the project within a four-year program of teaching and learning mathematics for social justice. We appreciate that the political aim of students using mathematics to develop an awareness of common issues for both communities is difficult to achieve (although we also note that the whole school is making its way toward social justice pedagogy and curriculum). First, the way CPS altered the originally planned school boundaries was something we had to contend with—that is, the historical tensions were reignited and in the air. Second, the local politician exacerbated these by pitting the neighborhoods against each other and proposing that the schools serve only Little Village students. Third, the politics of the immigration rights movement and the huge immigration marches nationally and in Chicago (where close to a million people participated in two large demonstrations) interacted with the specific conditions in the school campus in which African American students reported (to African American staff) that they did not fully feel their place in the building.

The opportunity is there to work with students to deconstruct and politically explore this polarized context, but existing contradictions can impede the process. For example, only 5 of about 30 African American Sojo students attended the May 1, 2006 pro-immigrant rights rally in Chicago (the larger of the two). I ran into an African American friend at the march who felt uncomfortable with two of the ubiquitous, mass-produced signs at the rally: "We Are All Immigrants" and "Immigrants Built America," neither of which is historically accurate and both of which negate the presence, contributions, experiences, and exploitation of both African Americans and Native Americans. There is a racially coded subtext here that is visible in the school and larger society both, with respect to "good" and "not-so-good" "minorities." Chicago employers report that they prefer hiring Mexican workers to African American ones because they were supposedly more "compliant" (Lipman, 2002). When asked about popular perceptions in their community about African Americans, Latina/o students report the stereotype that "Black people are lazy," while some African American students suggest that Mexican workers are "taking our jobs." A recent New York Times article (Swarns, 2006) conveyed these misconceptions well. In a Southern U.S. town in the state of Georgia, where Africans and African

Americans created most of the wealth and toiled mightily for centuries either as slaves or low-paid, exploited workers, a 51-year old Mexican worker was quoted as saying:

> They don't like to work, and they're always in jail. If there's hard work to be done, the blacks, they leave and they don't come back. That's why the bosses prefer Mexicans and why there are so many Mexicans working in the factories here.

The point here is that community knowledge *is* affected by popular misconceptions and myths.

Although this project had its limitations (Gutstein, 2006b), a strength was that we were able to tap into and build on students' community knowledge, and students were able to develop some critical and classical mathematical knowledge. The experience gives us (and others) some insight into the challenges and possibilities of teaching mathematics for social justice, although this was not a case in which we consciously investigated students' community knowledge. Rather, the generative theme emerged because of the dynamics of the situation. We might have ignored students' realities and kept to the already planned curriculum. Our analysis is that to have done so would have been a mistake and a missed opportunity to engage students and provide them a chance in school to examine their own lived experiences, deepen their sociopolitical awareness, and learn mathematics. One positive outcome we point to is that involving students in this particular project played a role in enculturating students to social justice pedagogy and reshaping their views of mathematics; their journaling after the project provided evidence for this assertion.

CONCLUSION

In the current school year (2006-07), our mathematics team has begun planning a more indepth, extended unit centered around *displacement* in an attempt to build on a generative theme salient for both communities. The specific local and broader national contexts shape our understanding of displacement. First, gentrification is a major issue in Chicago. While it affects many urban areas in the U.S., it is particularly severe here because the city power structure (i.e., Mayor Daley and his administration, major finance capitalists, and the real-estate/development machine) is in the throes of attempting to reshape Chicago as a global city (Lipman, 2004). The mayor and the school board are currently in the process of closing 60–70 neighborhood schools and creating 100 "new" ones, most of which are in the same school buildings but with large infusions of resources historically denied in the past (Lipman & Haines, in press). Many of the communities experiencing school closings are being rapidly gentrified. North Lawndale is very much on the list of affected neighborhoods, and has been referred to as "ground zero" by activists battling the redevelopment although the amount of new construction (e.g., condos) is still relatively small as of this writing. Thus displacement in the North Lawndale context refers to the on-coming gentrification in the community. Second, in Little Village, displace-ment refers to the removal of people out of the country altogether, back to Mexico. The U.S. House of Representatives passed a bill in September 2006 to build a 700-mile fence along the Mexican-U.S. border, and shortly afterwards, the Senate began considering the fence as well. In a small town of 37,000 located about 40 miles from Chicago, in early October 2006, town officials proposed an ordinance to penalize landlords who rented to un-documented immigrants and employers who hired them. Three thousand people showed up at the Town Hall in protest. Many residents of Little Vil-lage are undocumented, and the threat of expulsion from the community and country altogether is quite real. Thus both communities are faced with issues of displacement.

An appropriate challenge which we pose to ourselves is how do we know that this matters to students and community members, that this is really a generative theme when we have not done (for example) the thorough in-vestigation conducted by Brazilian teachers to uncover community knowl-edge? In October, 2006, we conducted focus group discussions and in-class discussions with small groups of students to explore this. In our conversa-tions with close to 60% of the sophomore class, students overwhelmingly expressed support and interest in the proposed unit. We also know, by the strength of the social movements for immigrant rights and against gentri-fication, that these issues matter profoundly to people (both adults and youth) in the affected communities. The tremendous number of people in the streets in support of immigrants and their rights is powerful evidence of this, and while the struggle against gentrification involves far fewer people, the level of consciousness and determination in impacted neighborhoods is quite high (Lipman & Haines, in press). We can *read the world* (Freire & Macedo, 1987) and understand clearly that the issue of displacement has deep meaning in Chicago.

While we have sketched out a political framework for this project, and have some clarity on how the community and critical knowledge fit in, there are certainly multiple challenges ahead of us. A key one is the con-nection of classical knowledge. The mathematics of change is central in understanding displacement in North Lawndale and Little Village. Spe-cific issues we plan to have students investigate include the changing de-mographics of the communities, the change in the cost and availability of properties, and the issues of affordability for people in the area. We want students to analyze the trends and the possibilities, as well as to think about possible actions to take, in conjunction with activists in their communities. We know from other gentrifying Chicago neighborhoods that the battle to stay in the area is an extremely difficult one, but there are community development corporations that are building or rehabilitating housing that is fairly affordable to many existing residents. This also entails mathematical analysis. Finally, we plan on having students investigate the mathematics of home ownership, loans, mortgages, and development schemes so that they begin to understand how capitalism works, and how real estate developers and banks profit while communities such as theirs experience extreme economic poverty and dis- and under-investment in basic human needs. All this will equip them with knowledge they will need as they become adults and have to fight to maintain their place in the neighborhood, city, and country. This, ultimately, is the goal of teaching (mathematics) for social justice— that students become agents of social change and join in, and eventually lead, the struggles to remake our world for peace and justice.

ACKNOWLEDGEMENT

Although this article is single authored, the teaching, planning, assessment, and analysis of the boundaries project in this story was collectively done with three other people besides the author: Joyce Sia (teacher), Phi Pham (teacher), and Patricia Buenrostro (mathematics support staff).

The research described here was partially supported by a grant from the National Science Foundation to the Center for the Mathematics Education of Latinos (No. ESI-0424983). The findings and opinions expressed here are those of the author and do not necessarily reflect the views of the funding agency.

NOTES

1. "We" refers to the school's two mathematics teachers (Phi Pham and Joyce Sia) and the other mathematics support staffperson (Patricia Buenrostro). Together, we constituted the school "mathematics team."
2. In brief, *generative themes* are key social contradictions in people's lives and the ways in which they understand them.

REFERENCES

Apple, M. W. (2004). Ideology and curriculum (3rd ed.). New York: Routledge Falmer.

Brantlinger, A. (2006). *Geometries of inequality: Teaching and researching critical mathematics in a low-income urban high school.* Unpublished doctoral dissertation. Evanston, IL: Northwestern University.

Brantlinger, A., Buenrostro, P., Gutstein, E., & Mukhopadhyay, S. (2007, March). *Teaching mathematics for social justice: Is the math there?* Presentation to be given at the annual meeting of the research presession of the National Council of Teachers of Mathematics, Atlanta.

Delpit, L. (1988). The silenced dialogue: Power and pedagogy in educating other people's children. *Harvard Educational Review, 58,* 280–298.

Fennema, E., & Scott Nelson, B. (1999). *Mathematics teachers in transition.* Mahwah, NJ: Erlbaum Associates.

Frankenstein, M. (1987). Critical mathematics education: An application of Paulo Freire's epistemology. In I. Shor (Ed.), *Freire for the classroom: A sourcebook for liberatory teaching* (pp. 180–210). Portsmouth, NH: Boyton/Cook.

Frankenstein, M. (1998). Reading the world with math: Goals for a criticalmathematical literacy curriculum. In E. Lee, D. Menkart, & M. Okazawa-Rey (Eds.), *Beyond heroes and holidays: A practical guide to K–12 anti-racist, multicultural education and staff development* (pp. 306–313). Washington D.C.: Network of Educators on the Americas.

Frankenstein, M., & Powell, A. B. (1994). Toward liberatory mathematics: Paulo Freire's epistemology and ethnomathematics. In P. L. McLaren & C. Lankshear (Eds.), Politics of liberation: Paths from Freire (pp. 74–99). New York: Routledge.

Freire, P. (1970/1998). *Pedagogy of the oppressed.* (M. B. Ramos, Trans.). New York: Continuum.

Freire, P. (1973). *Education for critical consciousness.* (M. B. Ramos, Trans.). New York. The Seabury Press.

Freire, P. (1994). *Pedagogy of hope: Reliving* Pedagogy of the Oppressed. (R. R. Barr, Trans.). New York: Continuum.

Freire, P., & Macedo, D. (1987). *Literacy: Reading the word and the world.* Westport, CT: Bergin & Garvey.

Gandin, L. A. (2002). *Democratizing access, governance, and knowledge: The struggle for educational alternatives in Porto Alegre, Brazil.* Unpublished doctoral dissertation. Madison, WI: University of Wisconsin.

Giroux, H. A. (1983). *Theory and resistance in education: Towards a pedagogy for the opposition.* Westport, CN: Bergin & Garvey.

Gutstein, E. (2006a). Building political relationships with students: An aspect of social justice pedagogy. To appear in (E. de Freitas & K. Nolan, Eds.), *Opening the research text: Critical insights and in(ter)ventions into mathematics education.*

Gutstein, E. (2006b). *Developing social justice mathematics curriculum from students' realities: A case of a Chicago public school.* Manuscript submitted for publication.

Gutstein, E. (2006c). *Reading and writing the world with mathematics: Toward a pedagogy for social justice.* New York: Routledge.

Gutstein, E., & Peterson, B. (Eds.). (2005). *Rethinking mathematics: Teaching social justice by the numbers.* Milwaukee, WI: Rethinking Schools, Ltd.

Hill, H., & Ball, D. L. (2004). Learning mathematics for teaching: Results from California's mathematics professional development institutes. *Journal for Research in Mathematics Education, 35*, 330–35.

Joseph, G. G. (1997). Foundations of Eurocentrism in mathematics. In A. B. Powell & M. Frankenstein (Eds.), Ethnomathematics: Challenging Eurocentrism in mathematics education (pp.61–81). Albany, NY: SUNY Press.

Knijnik, G. (1997). Popular knowledge and academic knowledge in the Brasilian peasants' struggle for land. *Educational Action Research, 5*, 501–511.

Ladson-Billings, G. (1994). *The dreamkeepers.* San Francisco: Jossey Bass.

Ladson-Billings, G. (1995a). Making mathematics meaningful in multicultural contexts. In W. G. Secada, E. Fennema, & L. B. Adajian (Eds.), *New directions for equity in mathematics education* (pp. 126–145). Cambridge: Cambridge University Press.

Ladson-Billings, G. (1995b). Toward a theory of culturally relevant pedagogy. *American Educational Research Journal, 32*, 465–491.

Lipman, P. (2004). *High stakes education: Inequality, globalization, and urban school reform.* New York: Routledge.

Lipman, P., & Haines, N. (in press). From accountability to privatization and African American exclusion – The case of Chicago public schools. *Educational Policy.*

Lorde, A. (1984). *Sister outsider: Essays and speeches.* Freedom, CA: Crossing Press.

Macedo, D. (1994). *Literacies of power: What Americans are not allowed to know.* Boulder, CO: Westview.

Mack, N. (1990). Learning fractions with understanding: Building on informal knowledge. *Journal for Research in Mathematics Education, 21*, 16–32.

McLaren, P. (2007). *Life in schools: An introduction to critical pedagogy in the foundations of education* (5th Edition). Boston: Pearson/Allyn and Bacon.

Moll, L. C., Amanti, C., & González, N. (Eds.) (2005). *Funds of knowledge: Theorizing practices in households, communities, and classrooms.* Mahwah, NJ: Erlbaum.

Tajitsu Nash, P., & Ireland, E. (1998). Rethinking terms. In B. Bigelow & B. Peterson (Eds.), *Rethinking Columbus: The next 500 years* (p. 112). Milwaukee, WI: Rethinking Schools, Ltd.

National Center for Research in Mathematical Sciences Education & Freudenthal Institute. (1997–1998). *Mathematics in context: A connected curriculum for grades 5–8.* Chicago: Encyclopedia Britannica Educational Corporation.

Osler, J. (2006). *Radical math website.* http://radicalmath.org/

Russo, A. (2003, June). Constructing a new school. *Catalyst.* Retrieved March 3, 2004 from http://www.catalyst-chicago.org/06-03/0603littlevillage.htm.

Shulman, L. S. (1986). Those who understand: Knowledge growth in teaching. *Educational Researcher, 15*, 4–14.

Skovsmose, O. (1994). *Towards a philosophy of critical mathematical education.* Boston: Kluwer Academic Publishers.

Skovsmose, O. (2004). *Critical mathematics education for the future.* Aalborg, Denmark: Aalborg University, Department of Education and Learning.

Swarns, R. (October 3, 2006). A racial rift that isn't black and white. *New York Times.* Retrieved October 4, 2006 from http://www.nytimes.com/

Tate, W. F. (1995). Returning to the root: A culturally relevant approach to mathematics pedagogy. *Theory into Practice, 34,* 166–173.

Turner, E. (2003). *Critical mathematical agency: Urban middle school students engage in mathematics to investigate, critique, and act upon their world.* Unpublished doctoral dissertation. Austin, TX: University of Texas—Austin.

CHAPTER 11

HOW MANY DEATHS?

Education for Statistical Empathy

Swapna Mukhopadhyay and Brian Greer
Portland State University, USA

ABSTRACT

In this paper, we suggest the term "statistical empathy" for the ability to relate
statistical data to the reality of what they stand for. To put the argument in stark
terms, we use historical and contemporary examples of representations of mass
killings. Alongside visual and literary artistic expressions, we exemplify mathematical tools designed to help convey the scale of such tragedies. We illustrate
the political processes of managing information through analysis of two highly
disputed issues, namely gun violence in the United States and the estimating of
excess civilian deaths in Iraq attributable to the American invasion.

INTRODUCTION

Yes, 'n' how many deaths will it take till he knows
That too many people have died?

—Bob Dylan (1962)

International Perspectives on Social Justice in Mathematics Education, pages 169–189
Copyright © 2008 by Information Age Publishing

> *It has been said that the mark of a truly educated person*
> *is to be deeply moved by statistics.*
>
> —Bill Moyers (2006)
> (quotation attributed to George Bernard Shaw)

Throughout history, including the present, there have been mass killings. In this paper, we discuss the mathematical procedures involved in counting, recording, or estimating death counts, analyzing the data statistically, and the mathematical and artistic means of representing them for the purpose of making an argument or conveying a sense of tragedy.

As we write, the media are reporting the loss of another 105 US military personnel in Iraq for the month of October, 2006. The losses on the Iraqi side—civilian and military—remain largely unmentioned in the US. There is no doubt that the world we live in is getting progressively more violent. The loss of life is often a result of intentional actions, as in a war or other sectarian violence. Besides the macro impact of intentional losses resulting from large-scale conflicts such as wars, there are innumerable instances of intentional killings that result from other conflictual interactions within societies, such as fighting among gangs.

Discussions on death and dying as a part of human violence may be characterized as morbid but they relate to a stark part of our reality that we cannot deny and should not ignore. Many people, in their post-modern busy life largely spent engaged in multitasking, may (rightfully) point out that there is little time for discussion or reflection. Some argue that too much exposure to grim realities numbs our sensitivity. Others evade the responsibility for action by saying "I am just a single individual, what can do I alone!" Or even "I do not do numbers!" Many of these responses stem from the general apathy characterized by the "not in my backyard" attitude, and by an alienation from mathematics. In this article, our goal is to provoke a conversation on contentious issues of contemporary life conveyed by numbers on which mathematics can be employed as a tool. We contend with Frankenstein (in press) that "we do need to know the meaning of the numbers describing our realities in order to deepen our understandings of our world." Following the quote from Shaw used by Bill Moyers, we call such understanding "statistical empathy."

In the National Governors' Summit in 2005, Bill Gates (2005) defined anew the three R's, the basic building blocks of better schools, as *rigor, relevance,* and *relationships*. As we reflect on this simple and yet profound framework, we ponder how mathematics as a school subject is viewed. For people involved in mathematics education, it is a painful truth that a vast number of people do not have a favorable view of mathematics as a domain of knowledge. School mathematics is typically considered to be boring, irrelevant, and meaningless, by adults and children alike.

One of us [SM] is a mathematics educator working with predominantly elementary school teachers. I often hear from my students that mathematics is an important subject but they fail to provide convincing examples of where mathematics is important in our everyday lives. They say that they use mathematics almost every day for balancing their checkbooks, and also when cooking. Some point out also that they use mathematics when making purchases and in estimating their daily travel times. In these conversations, the procedure for solving quadratic equations, or the proof of the theorem attributed to Pythagoras do not make an appearance. Conversely, it is arguable that most of the knowledge needed to negotiate the everyday situations raised by the students is gained out of school. (Note that we are not negating other motivations for teaching/learning mathematics, just pointing out that importance for mundane everyday functioning is a weak justification.) As one final example, consider the teaching of fractions, a staple of elementary school. To stimulate discussion, I say things like: "Do we really need to learn any fractions other than 1/2, 1/4, 3/4, 1/3, and 2/3? No recipe I have ever seen refers to 2/17 of a cup, for example." This is generally followed by an uncomfortable silence.

The point of sharing these anecdotes is to ask how we can start a sustaining conversation on the role of mathematics in the lives of our students at a deeper level. To exemplify how mathematics can help us expand our understanding of social and political issues that impact people, we deal in this paper with literally life and death issues. Although it might be argued that the social realities that we address in these examples are morbid and depressing, and should therefore be avoided, we would say that these examples are critical for youth—middle and high school students—who are often the target, and sometimes the perpetrators, of brutal violence. By engaging them in these conversations we hope to help the students become critically aware of the socio-political ramifications of violence and death. This conception of (mathematics) education is rooted in principles of democracy and social justice (Hackman, 2005; Mukhopadhyay & Greer, 2002) whereby we hope that the students can be brought to the realization that they are capable of framing and voicing their opinions, and acting on them, based on critical thinking rather than remaining mere passive consumers of information. An additional advantage of this awakening of agency, we believe, is the way it can affect perception of mathematics. From characterizing mathematics as "boring and useless," "hard," "I-am-not-good-at-it," students may start valuing mathematics as an essential and powerful tool for, in Freire's phrase, reading and writing the world.

Who Gets Killed?

On the twentieth anniversary of the death of John Lennon, his widow, the artist Yoko Ono, paid for the display of large posters in New York, Los Angeles and Cleveland, Ohio showing a photograph of blood-stained glasses and the New York skyline with herself in front, bearing the text "Over 676,000 people have been killed by guns in the U.S.A. since John Lennon was shot and killed on December 8, 1980." (Berger, 2001; BBC, 2000; Gurney, n.d.). Later, Ono commented, "The number of people who have died by gunshot since John's death is 10 times larger than the total number of American soldiers lost in the Vietnam War. It's like we are living in a war zone."

The starkly simple statement on the poster may have caught many bypassers' attention. Maybe the billboard was noticed because of John Lennon's celebrity status. But does the number in the statement provoke reflection and analysis? 676,000 people killed by guns in 20 years readily tells us that if the trend continues, over 30,000 people will die by gun deaths every year. Further computation converts this to about 93 gun deaths per day, roughly equivalent to 4 deaths per hour. In the most concrete terms, this equates to about one death every fifteen minutes.

The most recent published data from the U.S. Centers for Disease Control and Prevention show that 2,827 children and teens died from gunfire in 2003. This figure amounts to "...one child or teen every three hours, nearly eight every day, 54 children and teens every week" (Children's Defence Fund, 2006, 1). By contrast: "The number of children and teens killed by gun violence in 2003 alone exceeds the number of American fighting men and women killed in hostile action in Iraq from 2003 to April 2006" (CDF, 2006, p. 2).

Although the latest data show a slightly downward trend (CDF, 2005), we learn a set of very alarming facts:

- The number of children and teens in America killed by guns in 2003 would fill 113 public school classrooms of 25 students each.
- The number of children and teens in America killed by guns since 1979 would fill 3,943 public school classrooms of 25 students each.
- Almost 90 percent of the children and teens killed by firearms in 2003 were boys.
- The firearm death for Black males ages 15 to 19 is more than four times that of White males the same age.
- A Black male has a 1 in 72 chance of being killed by a firearm before his 30th birthday.
- A White male has a 1 in 344 chance of being killed by a firearm before his 30th birthday (CDF, 2006. p. 2).

The cost of gun violence to society, through injuries as well as deaths, and the consequent extensive medical and social care, is an additional burden to the tax-payer. Admitting that precision in information is difficult to achieve, Cook and Ludwig (2002, p. 97) point out that:

...the national costs of gun violence are roughly $100 billion per year, with $15 billion or more attributable to gun violence against youth. The tangible costs to the victims from medical expenses and lost productivity are only a small part of the overall problem. The real burden of gun violence comes from the cost of public and private efforts to reduce the risks, and the fear of victimization that remains despite these efforts.

Kids in the Line of Fire (VPC, 2001) provides an in-depth analysis of the link between children, handguns, and homicide. It is based on analysis of homicide data for the five-year period 1995 through 1999. (For a comprehensive timeline of worldwide gun violence in schools, see http://www.infoplease.com/ipa/A0777958.html)

The Violence Policy Center (2001) has problematized the assumption that gun violence among youth is primarily an inner-city problem associated with criminals, gangs, and people of color:

Recent school shootings have garnered greater publicity than in previous years, with one clear reason being the larger number of victims. Perhaps just as important is the demographic profile of the victims and shooters: mostly white, from either the suburbs or rural America. As a result of the high rates of violence seen among urban, primarily black, youths in the late 1980s and early 1990s, such violence came to be seen by many as solely a plague of the cities. Viewing the issue literally in terms of black and white, rural, white youth were portrayed as having "respect" for guns, using them only for hunting or other sporting activities. Shootings among black youths were often falsely portrayed as a virtually inevitable, almost normal, component of the urban environment. And when "good" kids go bad, the gun lobby is quick to blame virtually anything—television, movies, bad parenting, even an undefined "wave of evil"—except the one thing that comes up time and time again: the easy availability of handguns. (Retrieved 10/31/06 from: http://www.vpc.org/studies/wgunint.htm)

On the other hand, the National Rifle Association, a very powerful group that lobbies for gun ownership, describes the Violence Policy Center as "the most effective ... anti-gun rabble rouser in Washington." We cite this statement here to illustrate the degree to which debate on gun control is polarized and politicized, leading to "furious politics, marginal policy" (Spitzer, 1998, p. 133). An interesting format reflecting this polarity is used in Haerens (2006) in which chapters are presented pair-wise, with one author in each pair on each side of the controversy. For example, a chapter entitled "Youth gun violence is a serious problem" is followed by another entitled "The problem of youth gun violence is exaggerated."

A fundamental question is: To what extent can such questions be decided by research? A review sponsored by the National Research Council concluded that: "While there is a large body of empirical research on firearms and violence, there is little consensus on even the basic facts about these important policy issues" (Wellford, Pepper, & Petrie, 2004, p. 1). Nevertheless, the report, in making a number of recommendations for future research, implies that properly done research could settle questions. Thus, at one point it is stated that: "Ultimately, it is an empirical question whether defensive gun use and concealed weapons laws generate net social benefits or net social costs" (p. 6). We believe that is missing a crucial point. No matter how well done the research, it will not settle the question since opinions will differ on how social benefits and costs are to be measured, which is a question of values.

In this context, we can illustrate our own praxis in mathematics education—putting ideology into action. Education for social justice, for us, is to encourage learners to actively participate in their own education so that they, with their teachers, identify and acknowledge issues of injustice and devise an action plan. For example, statistics education as data handling has been a part of school mathematics curriculum for a while but it seldom creates a context where the students are empowered to act on an issue that they are studying together. In lessons on statistics, children tabulate and graph modes of transportations to school, favorite cereal, and suchlike, hardly ever spending any discussion on the social context. In many cases, the successful students produce picture-perfect graphs without being able to articulate the deep underlying implications. Classroom instruction often emphasizes the procedure and the right answer without getting into the sense-making aspects of mathematics—a practice that is exacerbated by the current culture of standardized testing. Papert (1993), critiquing a similar practice of early and massive imposition on children of "letteracy," an impoverished form of literacy consisting merely of the ability to decode strings of alphabetic letters. By analogy with Freire's term for this activity, "barking at words," mathematics should not be reduced to pawing at symbols. Likewise, following Freire's emphasis on reading the word and the wor(l)d, we recommend moving mathematics instruction from mere symbol manipulation to the development of a sense of critical thinking.

Heather Hackman (2005), quoting Bell, points out that social justice education is both the goal and the process. Thus, "the process for attaining the goal of social justice should be democratic and participatory, inclusive and affirming human agency and human capacities for working collaboratively to create change" (p. 104).

Echoing Hackman, who identifies five essential components of social justice education: content mastery, tools for critical analysis, tools for personal reaction, tools for action and social change, and an awareness of multicultural group dynamics (p. 104), we present an approach to address social justice within the context of mathematization of the real world of violent death in Figure 11.1.

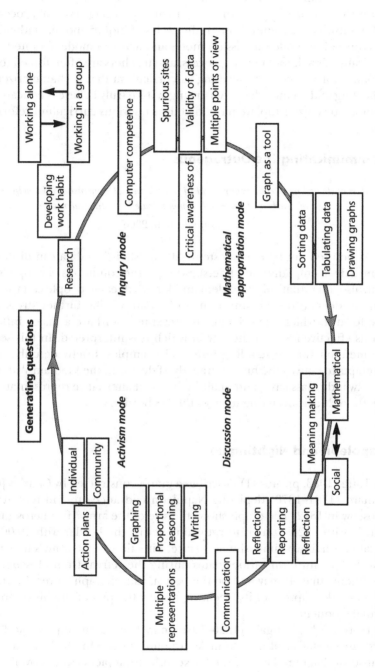

Figure 11.1   Cycles of discussions to foster statistical empathy.

For us, from the perspective of teaching, generating questions for a situation (here, issues of gun violence) entails a four-stage cyclical process that refines itself as one goes through the cycles of inquiry mode, mathematical appropriation mode, discussion mode and activism mode. Granted, pedagogically it is a time-intensive process, and teachers are often forced to compromise for the sake of "covering" a curriculum that is characterized as a mile long and an inch deep. Accordingly, if we truly believe is educating for democratic citizenship, we have to embrace serious curriculum reform.

Communicating the Outrageous

Part of struggling to change our world in the direction of more justice is knowing how to clearly and powerfully communicate the outrageousness.
—Frankenstein, 2006

As pointed out by Frankenstein (2006) "statistical data can distance us from a deep empathy and understanding of the conditions of people's lives. But, also, quantitatively confident and knowledgeable people can use those data to deepen their connections to humanity." She further stresses that the form in which the evidence is presented can have a major influence on its effectiveness and the way in which it is interpreted. In this section, we mention two outstanding historical examples, briefly describe artistic attempts to convey the human tragedy of deaths in the Vietnam War and in the Rwanda massacres, and finally we present alternative representations of deaths of Mexicans trying to cross the US border.

Napoleon and Nightingale

Tufte (1983, pp. 40–41) describes a graph constructed by Charles Joseph Minard (1781–1870) depicting Napoleon's advance on, and retreat from, Moscow in 1812. The graph shows the size of the army (from crossing into Russia with 422,000 men to returning across the border with 10,000), its location on a two-dimensional surface, direction of the army's movement, and temperature on various dates during the retreat from Moscow. Tufte comments that "it may well be the best statistical graphic ever drawn" and quotes a description of its "seeming to defy the pen of the historian by its brutal eloquence."

Florence Nightingale (1820–1910) may have been responsible for the first use of statistical data to make a political case. In 1858, she devised a statistical diagram, labeled the "coxcomb" to depict changes over time in deaths, on the battlefield and in the hospitals, of British soldiers in the

Crimean War. With this and other statistical data, she managed to convince the army authorities to make major changes in policy that resulted in substantial decreases in hospital deaths. The diagram may be viewed at www.florence-nightingale-avenging-angel.co.uk/Coxcomb.htm

The accompanying text explains:

> The Government would not allow her to publish her most damning statistics which showed that hospital conditions were the main cause of death. In this published diagram, therefore, she tried to support her case for better hygiene by using published Army figures to show that the death rate decreased after the Sanitary Commissioners cleaned up the hospitals. Her opponents claimed that the reduction in death rate resulted from other changes that occurred at the same time.

Vietnam and Rwanda

A powerful piece by Frankenstein (2006) drew our attention to these examples, and should be consulted for more detail. The Vietnam Veterans Memorial in Washington, D. C., bearing the names of 57,939 Americans who died in that war, is well known (see, e.g. a description by Tufte (1990, p. 43) of its design effectiveness so that "we focus on the tragic information"). Less well known is *The other Vietnam Memorial* by artist Chris Burden, USA):

> In this work, Burden etched 3,000,000 names onto a monumental structure that resembles a Rolodex standing on its end. These names represent the approximate number of Vietnamese people killed during U.S. involvement in the Vietnam War, many of whom are unknown. Burden reconstructed a symbolic record of their deaths by generating variations of 4000 names taken from Vietnamese telephone books. By using the form of a common desktop object used to organize professional and social contacts, Burden makes a pointed statement about the unrecognized loss of Vietnamese lives." (notes from the Museum of Contemporary Art in Chicago, IL, cited by Frankenstein, 2006).

Artist Alfredo Jaar (born in Chile, works in New York City) went to Rwanda in 1994 to try to understand and represent the slaughter of "possibly a million Tutsis and moderate Hutus" during three months of Prime Minister Jean Kambanda's term.

> Even after 3000 [photographic] images, Jaar considered the tragedy to be unrepresentable. He found it necessary to speak with the people, recording their feelings, words and ideas....In Jaar's Galerie Lelong installation, a table containing a million slides is the repetition of a single image, The Eyes of Gutete Emerita." The text about her reads: "Gutete Emerita, 30 years old, is standing in front of the church. Dressed in modest, worn clothing, her hair is

hidden in a faded pink cotton kerchief. She was attending mass in the church when the massacre began. Killed with machetes in front of her eyes were her husband Tito Kahinamura (40), and her two sons Muhoza (10) and Matrii-gari (7). Somehow, she managed to escape with her daughter Marie-Louise Unumararunga (12), and hid in the swamp for 3 weeks, only coming out at night for food. When she speaks about her lost family, she gestures to corpses on the ground, rotting in the African sun.

The art review ends with a comment about the numbers: " The statistical remoteness of the number 1,000,000 acquires an objective presence, and through the eyes of Gutete Emerita, we witness the deaths, one by one, as single personal occurrences" (Rockwell, 1998).

Border Crossings

A large number of people have died since *Operation Gatekeeper*, a program to seal the U.S.–Mexico border, was introduced in 1994 (Fig. 11.2).

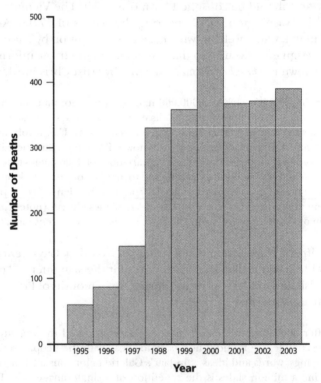

Figure 11.2 Number of deaths crossing U.S.–Mexico border around San Diego, CA, 1995–2003.

In February, 2005, one of us [SM] participated with a group of educators in a study tour organized by Global Exchange in the San Diego-Tijuana area on issues of border-crossing. (Bigelow, 2006). The outside walls of Tijuana airport, we noticed, were covered with elaborate "border art" depicting the violence and deaths that many ordinary Mexicans face as they attempt to cross the border illegally. Makeshift memorials were created as simple crosses with the names, age and origins of the dead (Fig. 11.3). Moreover, besides the individual crosses, there are displays of colorful and elaborately designed coffins, carrying the numbers of deaths, one for each year from 1995 to 2003 (Fig. 11.4).

Figure 11.3 Tijuana Airport, 2005: crosses (© Mukhopadhyay, 2005)

Figure 11.4 Tijuana Airport, 2005: coffins (© Mukhopadhyay, 2005)

These examples, as well as those described above, illustrate the power of powerful representations to mediate between bare statistics, numbers on paper, and the reality for which those statistics are referents.

Such representations may be uncomfortable. In February, 2003, David Cohen reported on Slate that:

> Earlier this week, U.N. officials hung a blue curtain over a tapestry reproduction of Picasso's *Guernica* at the entrance of the Security Council. The spot is where diplomats and others make statements to the press, and ostensibly officials thought it would be inappropriate for Colin Powell to speak about war in Iraq with the 20th century's most iconic protest against the inhumanity of war as his backdrop.

How Many Iraqi Deaths?

> *Death has a tendency to encourage a depressing view of war.*
> —Donald Rumsfeld

On October, 2006, the British medical journal, The Lancet, published a paper (Burnham, Lafta, Doocy, & Roberts, 2006) in which the authors reported that:

> We estimate that as of July, 2006, there have been 654,965 (392,979–942,636) excess Iraqi deaths as a consequence of the war, which corresponds to 2.5% of the population in the study area. (Retrieved 10/30/06 from http://www.thelancet.com/webfiles/images/journals/lancet/s0140673606694919.pdf

Note that 654,965 is a point estimate, the number that the analysis identifies as the single most probable; the numbers in brackets are those for a 95% confidence interval. A longer report (Burnham, Doocy, Dzeng, Lafta, & Roberts, 2006) provides more detail and context. In a similar study carried out in 2004 (Roberts, Lafta, Garfield, Khudhairi, & Burnham, 2004) the estimate of excess mortality during the 17.8 months after the 2003 invasion was 98,000, with a 95% confidence interval of 8000–194000 (excluding the data from Fallujah).

In both cases, the estimates were very much higher than others obtained using different methodologies, have been widely contested in the media, and dismissed as not credible by government leaders in the US and UK, and, in the more recent case, Iraq and Australia. For example, President Bush, questioned by Suzanne Malveaux of CNN at a White House Press Conference said that he did not consider the report credible, that the methodology had been "pretty well discredited" and that he stood by the number 30,000 that he had cited previously. He referred to the estimate in the Lancet report as "600,000, or whatever they guessed at" (Retrieved 10/30/06 from www.whitehouse.gov). No further questions were asked on this topic during the press conference. Nevertheless, the President's statement was very widely quoted in the media, often in headlines. Richard Garfield, a public health professor at Columbia University who works closely with a number of the authors of the report commented:

> I loved when President Bush said "their methodology has been pretty well discredited." That's exactly wrong. There is no discrediting of this methodology. I don't think there's anyone who's been involved in mortality research who thinks there's a better way to do it in unsecured areas. I have never heard of any argument in this field that says there's a better way to do it. (Murphy, 2006)

The appeal to experts by journalists is deserving of analysis. Typically, articles following the publication of the report cite short comments by a number of such experts. The divergence in the opinions cited is typical of what happens when statistical experts give opinions on a complex study. It is the nature of statistical applications of this level of complexity that agreement is not to be expected. There are aspects of the methodology that represent potential weaknesses in the design—indeed, the authors themselves clearly identify and discuss several. To elevate such disagreement to a claim that the methodology has been discredited shows ignorance of the nature of statistical research.

There is an irony in that reporting on a study based on sampling, there is no mention of the samplings implicit in the above and similar examples of press coverage. First, the experts quoted are a sample—and probably what is technically called a "convenience sample," rather than one that is representative of the appropriate population. Secondly, the short quotations are, almost inevitably, sampled on the basis of the journalists' subjective criteria, from longer and more nuanced statements. Interactions with the experts are usually one-shot deals, with short quotations of what the experts say. It is not uncommon for the experts to want to clarify or correct statements attributed to them, but such an opportunity is rarely afforded. For example:

> The *Washington Post*, perhaps most damagingly to the study's reputation, quoted Marc E. Garlasco, a senior military analyst at Human Rights Watch, as saying, "These numbers seem to be inflated."
>
> Mr. Garlasco says now that he had not read the paper at the time and calls his quote in the *Post* "really unfortunate." He says he told the reporter, "I haven't read it. I haven't seen it. I don't know anything about it, so I shouldn't comment on it." But, Mr. Garlasco continues, "like any good journalist, he got me to."

Mr. Garlasco says he misunderstood the reporter's description of the paper's results. He did not understand that the paper's estimate includes deaths caused not only directly by violence but also by its offshoots: chaos leading to lack of sanitation and medical care. (Guterman, 2005)

An example of a more extended interaction is provided by the website Media Lens (www.medialens.org). A professor of mathematics, well known for his books, including *A Mathematician Reads the Newspaper* (Paulos, 1996) wrote in the British newspaper, *The Guardian* (Paulos, 2004):

Given the conditions in Iraq, the sample clusters were not only small, but sometimes not random either... So what's the real number? My personal assessment, and it's only that, is that the number is somewhat more than the IBC's confirmed total, but considerably less than the Lancet figure of 100,000.

After Media Lens commented that they "had not found a single example anywhere in the British or US press of a commentator rejecting estimates of 1.7 million deaths in Congo produced by the same lead researcher (Les Roberts) and offering their own "personal assessment" in this way," Paulos responded:

I regret making the comment in my Guardian piece that you cite... I still have a few questions about the study (moot now), but mentioning a largely baseless 'personal assessment' was cavalier. I should simply have stated my doubts about the study's scientific neutrality given what seemed at the time like an expedient rush to publish it."

John Allen Paulos Math Dept, Temple Univ" (Email to Media Lens, September 7, 2005, retrieved 11/05/06 from http://www.medialens.org/alerts/05/050906_burying_the_lancet_update.php)

The criticism that the timing of publication of the report was politically motivated (it came out shortly before the presidential election of 2004) is widespread. One of the authors of the report, Les Roberts, replied to Paulos giving the reasons for the timing of publication, the most important of which was his belief that if it had not come out until after the election, it would have been interpreted as a cover-up. Paulos then stated that "I understand now the situation surrounding the study's original publication." What is striking about this example is that an extended and logical debate led to some reasonable consensus. Another example of such an extended interchange facilitated by the Media Lens group included follow-up to a BBC programme whereby Les Roberts was given the opportunity to respond to listeners' questions.

Journalists, who cannot be expected to have the statistical expertise to evaluate technical reports, do not always take the obvious step of seeking expert advice. Again we turn to Media Lens for a fascinating and fully documented example (see the website for the full account). It began with an editorial that claimed that the Lancet findings had been reached "by extrapolating from a small sample... While never completely discredited, those figures were widely doubted" (Leader, 'The true measure of the US and British failure,' The Independent, July 20, 2005)

David Edwards of Media Lens challenged the Independent's Mary Dejevsky, senior leader writer on foreign affairs to indicate the basis for the claim that the sample was small. Dejevsky responded:

> ... personally, i think there was a problem with the extrapolation technique, because—while the sample may have been standard for that sort of thing—it seemed small from a lay perspective (i remember at the time) for the conclusions being drawn and there seemed too little account taken of the different levels of unrest in different regions. my main point, though, was less based on my impression than on the fact that this technique exposed the authors to the criticisms/dismissal that the govt duly made, and they had little to counter those criticisms with, bar the defence that their methods were standard for those sort of surveys. (Retrieved 10/30/06 from http://www.williambowles. info/media/2005/ml_lancet.html)

Edward Herman, co-author with Noam Chomsky of the classic media study, *Manufacturing Consent*, commented:

> Massive incompetence in support of a war-apologetic agenda. Dejevsky objects to the figures because they are vulnerable to discrediting for reasons that make no sense. I wonder if she finds sampling discreditable in all cases. (Email to Media Lens, September 1, 2005, retrieved 10/30/06 from http:// www.williambowles.info/media/2005/ml_lancet.html)

Media Lens commented that, operating on a shoestring, in allowing the right of reply and continuing the discussion they had performed a function almost totally abdicated by the media.

A simple search on the Internet will produce many commentaries on the two reports, in many cases predictable given the political stance of the authors and/or the publications (e.g., Hitchens, 2006; Moore, 2006). There is also abundant evidence of the relativization of opinion and democratization of ignorance that discussion groups on the Internet typically generate. We will restrict ourselves to a single example. Somebody contributed as follows:

> That Lancet study is poorly done. The actual range of estimated civilian deaths was something on the order of about 10,000–100,000. That is a wide range that lends NO credence to the 100,000 number being selected over

the 10,000 number. It was a politically biased article and never should have made it to print, at least in the form it was written. (Retrieved 10/31/06 from http://www.sport-groups.com/board/nextpost/93676/0)

After another contributor pointed out that the report gave a 95% confidence interval of 8,000–194,000 with a point estimate of 98,000 the original contributor persisted as follows:

I didn't bother to look it up because the range was so varied. My point was in a range so large there is no way to pick one number over the other. That the article was flawed is true and that it should not have been published is true. (Retrieved 10/31/06 from http://www.sport-groups.com/board/next-post/93676/0)

It is to be expected that almost all people (including us) will react to these studies in alignment with their existing political views. This reality fundamentally challenges the notion of conclusions being reached, at least partly, on the basis of scientific evidence. Many criticisms of the reports claim that the political views of the authors and of the editor of the Lancet discredit the data. At least in the case of Les Roberts, the authors are anti-war, more specifically opposed to the invasion and war in Iraq, as is the editor of the Lancet, Richard Horton (and as, indeed, are the authors of this paper). What are the implications? Are people with such views considered incapable of carrying out studies of this sort and having the findings taken seriously? Such a position rests on the myth of science and mathematics being value-free, ethically neutral, and apolitical.

It is worth remembering that *The Lancet* is one of the most highly respected scientific journals, and that papers published in it are subject to the most stringent peer review. Apparently, however, it should not deal with deaths in war when those deaths are caused by "us." A June 23, 2005 editorial in the Washington Times lamented what it saw as an instance of "egregious politicization of what is supposed to be an objective and scientific journal." Why is it unreasonable that a journal serving a profession whose members take an oath to protect human life should raise issues about the avoidable killing of human beings?

What Does This All Mean for Mathematics Education?

As we discussed in the introduction, a major reason for the alienation towards mathematics widespread in society is its lack of relevance to people's lives. Elsewhere (Mukhopadhyay & Greer, 2004, p. 201) we have proposed that a better slogan than "Mathematics for all" is "Mathematics of all":

> By [mathematics of all] we mean, on the one hand, recognizing the diversity
> of human activity that is mathematical, and on the other, promoting the idea
> of every individual being a person who can meaningfully use mathematics.

Mathematics could be taught as, amongst other things, a tool for making sense, and then acting upon, issues that for the students, and the adults they will become, are of importance to them personally, to their communities, and to society in general. Adapting Freire's phrase, this is a vision of teaching mathematics "for reading and writing the world" (Gutstein, 2006).

In the current political climate of the United States, we perceive a chronic lack of the analytical tools that mathematics education ought to equip people with, a manifestation of what Macedo (2000, p. 5) calls "education for stupidification." As Chomsky (2000, p. 24) stated: "The goal is to keep people from asking questions that matter about important issues that directly affect them and others." For too many students in the United States, these may literally be life-and-death issues.

It has been widely documented that people, in general, have a weak understanding of numerical data. In particular, it is difficult for many people to grasp the meaning of large numbers. Lack of numeracy is compounded by a lack of understanding of basic statistical principles such as sampling variation, randomness, margin of error (as is evident in our discussion above of the media treatment of the report on Iraqi deaths). For example, as we complete the writing of this chapter, the mid-term elections in the US are about to happen and the media are full of terms such as "statistical dead heat" which only a tiny fraction of the electorate understand. Shouldn't such understanding be part of what is considered an adequate mathematical education?

Beyond the basic statistical principles, there is the need for more sophisticated forms of argumentation. In the debate on gun control, it is common to see an argument (from either "side") formulated along these lines:

1. There are countries, such as Switzerland and Canada, where guns are readily accessible, yet homicide levels are low in those countries.
2. These facts prove that a high level of gun violence is not caused by access to guns.
3. Therefore attempts to reduce gun violence in the US by blocking access to guns will not work.

The conclusion may be true, but the argument is not valid. It rests on a simplistic assumption of a deterministic single cause/ single effect model, ignoring the clear reality that the causes of gun violence are multiple, and arguably deeply socio-cultural in ways that are extremely difficult, if not impossible, to quantify. We argue, accordingly, that mathematics education should convey some of the complexity of mathematically modeling social phenomena and a sense of what demarcates questions that can be answered by empirical evidence from those that depend on value systems and world-views. The interaction between statistics, politics, modeling of social phenomena, and views of people (e.g. Hacking, 1975; 1990) is a major part of the social history of mathematics and of modern political history, virtually absent from mathematics education at all levels.

Technological advances mean that the amount of information available is swamping people's intellectual and analytical tools for making sense of, and critically evaluating, opposing claims. An interesting approach is in the Opposing Viewpoints Series, the volume of which that deals with gun violence we referred to above. In the introduction (Haerens, 2006, p. 9) points out that "the more inundated we become with differing opinions and claims, the more essential it is to hone critical reading and thinking skills to evaluate these ideas."

A contributory factor to weak understanding of the evaluation of evidence in the modeling of social phenomena is inadequate cultural support, in particular from the media. In many cases, it is possible to go beyond vague assertions of bias through the use of relatively simple analysis. For example, in relation to our own local newspaper:

> From May-November 2004, 116 Palestinian children and 8 Israeli children were killed. The Oregonian reported all but one of the Israeli children's killings in a headline while reporting only 2 (under 2%) of the Palestinian children's deaths in headlines. (Retrieved 11/05/06 from http://www.auphr. org/oregonian.php)

Tools that search databases enable simple yet powerful analyses that lend support to assertions such as that the newspapers did not adequately report the 2006 Lancet paper. During the week following its publication, a search in LexisNexis found the following frequencies (Jack Straw, a British minister, had recently made statements that British Muslim women should not wear veils, and the pop star Madonna had just adopted an African baby):

Jack Straw + veil	348
Madonna + adoption	219
Iraq + Lancet	44

The issues that we have raised in this paper speak to central questions. Munir Fasheh (1997) asked the startling question "Is mathematics dead?." We interpret this to mean that mathematicians too often adopt a stance of neutrality, distancing their work from its impact on people's lives, and that mathematics education does not deal with students' lived experience. In this chapter, we have used the example of numeration of mass deaths to argue that mathematics could be brought back to life as a tool for communicating outrageousness and provoking outrage. As mathematics educators, we seek ways to expand statistical empathy through the imaginative reframing of numerical data, through ingenuity in the design of statistical diagrams and schematic representations, and through simulations (see Petersen & Gutstein, 2005, for several examples), standing alongside expression through the visual and literary arts, such as the work seen in Mexico, the Rwanda genocide portrayal, John Donne's famous passage that begins "No man is an island," Bob Dylan's song "Blowin' in the wind." Such attempts address the question posed so passionately by Ubiratan D'Ambrosio, in this volume and elsewhere, namely what are the ethical responsibilities of mathematicians and mathematics educators as we seek survival with dignity?

REFERENCES*

BBC News. (2000, December 6). *Yoko's anniversary peace call*. Retrieved 10/30/2006, from http://news.bbc.co.uk/2/hi/entertainment/1056493.stm

Berger, M. (2001). Times Square photo project. Retrieved 11/01/2006, from http://www.w42stnyc.com/stock02/words/gunskill.htm

Bigelow, B. (2006). *The line between us*. Milwaukee, WI: Rethinking Schools.

Burnham, G., Doocy, S., Dzeng, E., Lafta, R., & Roberts, L. (2006). *The human cost of the war in Iraq*. Retrieved 10/11/2006 from http://www.thelancet.com

Burnham, G., Lafta, R., Doocy, S, & Roberts, L. (October 21, 2006). Mortality after the 2003 invasion of Iraq: A cross-sectional cluster sample survey. *The Lancet*, 368(9546), 1421–1428.

Children's Defense Fund (2005). *Protect children not guns*. Retrieved 1/25/2005 from http://www/childrensdefense.org

Children's Defense Fund (2006). *Protect children not guns*. Retrieved 10/28/2006 from http://www.childrensdefense.org/site/DocServer/gunrptrevised06. pdf?docID=1761

CNN.com (2000, February 29). *First-grader shot dead by classmate in Michigan school*. Retrieved 11/02/2006, from http://archives.cnn.com/2000/US/02/29/school.shooting.03/

Cook, P. J., & Ludwig, J. (2002). The cost of gun violence against children. *The Future of Children*, 12(2), 87–99. Retrieved 11/02/06, from http://www.futureofchildren.org/pubs-info2825/pubs-info_show.htm?doc_id=154414

Dylan, B. (1962). *Blowin' in the wind*. Retrieved 11/01/2006, from http://www.bobdylan.com/songs/blowin.html

Fasheh, M. (1997). Is mathematics in the classroom neutral—or dead? *For the Learning of Mathematics*, 17(2), 24–27.

Frankenstein, M. (in press). Quantitative form in arguments. In J. Spring, H. J. Silverman & D. A. Gabbard (Ed.), *Knowledge and power in the global economy*.

---

* Most of the papers retrieved from the Internet have been archived by the authors as Word files, and are available on request.

Mahwah, NJ: Erlbaum. http://www.media.pdx.edu/Mukhopadhyay/Frankenstein_062206.asx

Gates, B. (2005). *National education summit on high schools* [Electronic Version]. Retrieved 10/31/2006 from http://www.admin.mtu.edu/ctlfd/Ed%20Psych%20Readings/BillGates.pdf

Gurney, S. (n.d.). In December 2000, *The Times.* Retrieved 10/01/2006, from http://a-i-u.net/antigun_b.html

Guterman, L. (2005, January 27). *Researchers who rushed into print a study of Iraqi civilian deaths now wonder why it was ignored.* Retrieved 10/31/2006, from http://chronicle.com/free/2005/01/2005012701n.htm

Gutstein, E., & Peterson, B. (Eds.) (2005). *Rethinking mathematics: Teaching social justice by the numbers.* Milwaukee, WI: Rethinking Schools.

Hacking, I. (1975). *The emergence of probability.* Cambridge: Cambridge University Press.

Hacking, I. (1990). *The taming of chance.* Cambridge: Cambridge University Press.

Hackman, H. (2005). Five essential components of social justice education. *Equity & Excellence, 38,* 103–109.

Haerens, M. (2006). *Gun violence.* Farmington Hills, MI: Greenhaven Press.

Hitchens, C. (2006, October 16). *Epidemiology meets moral idiocy.* Retrieved 10/31/2006, from http://www.slate.com/id/2151607/fr/rss/

Infoplease (n.d.). *A time line of recent worldwide school shootings.* Retrieved 10/31/2006 from http://www.infoplease.com/ipa/A0777958.html

Macedo, D. (1994). *Literacies of power: What Americans are not allowed to know.* Boulder, CO: Westview Press.

Macedo, D. (Ed.). (2000a). *Chomsky on MisEducation.* Lanham, MD: Rowman & Littlefield.

Macedo, D. (2000b). Introduction. In D. Macedo (Ed.) *Chomsky on MisEducation* (pp. 1–14). Lanham, MD: Rowman & Littlefield.

Moore, S. E. (2006, October 18). *655,000 War Dead? A bogus study on Iraq casualties.* Retrieved 10/31/2006, from http://www.opinionjournal.com/editorial/feature.html?id=110009108

Moyers, B. (2006). *America 101.* Retrieved 11/01/2006 from http://www.commondreams.org/views06/1101-33.htm

Mukhopadhyay, S. & Greer, B. (2002). Mathematics for socio-political criticism: The issue of gun violence. In S. C. Agarkar & V. D. Lale (Eds.), *CASTME-UNESCO-HBCSE International Conference on Science, Technology and Mathematics Education for Human Development* (Vol. 2, pp. 195–199). Goa, India: Homi Bhabha Centre for Science Education/Tata Institute of Fundamental Research.

Mukhopadhyay, S., & Greer, B. (2004). Teaching mathematics in our multicultural world. In A. M. Johns & M. K. Sipp (Eds.) *Diversity in College Classrooms: Practices for Today's Campuses* (pp. 187–206). Ann Arbor, MI: University of Michigan Press.

National Research Council (2005). *Firearms and violence: A critical review.* Washington. D.C.: The National Academy Press.

Papert, S. (1993). *Obsolete skill set: The 3 Rs.* Retrieved 11/04/2006, from http://www.wired.com/wired/archive/1.02/1.2_papert.html?topic=&topic_set

Paulos, J. A. (1996). *A mathematician reads the newspaper.* New York: Anchor Books.

Spitzer, R. J. (1998). *The politics of gun control* (2nd Ed.). New York: Chatham House Publishers.

Tufte, E. R. (1983). *The visual display of quantitative information.* Cheshire, CT: Graphics Press.

Tufte, E. R. (1990). *Envisioning Information.* Cheshire, CT: Graphics Press.

Violent Policy Center (April, 2001). *Where'd they get their guns?* Retrieved 11/02/2006 from http://www.vpc.org/studies/wguncont.htm11

CHAPTER 12

FUNDAMENTAL REASONS FOR MATHEMATICS EDUCATION IN ICELAND

Kristín Bjarnadóttir
Iceland University of Education, Reykjavík

ABSTRACT

Iceland was predominantly a rural society under Danish rule until the 20th century. The paper discusses arguments and reasons for the presence and absence of mathematics education in Iceland through the centuries. The content of mathematics education traditionally concerned trade and prerequisites for university entrance until the 1960s. At that time the OECD channelled to Iceland radical ideas about the content and purpose of mathematics education. At that moment and others, when mathematical education was at crossroads, arguments brought up by influential individuals, referring to fundamental reasons for mathematics education, were of a great importance. The pros and cons of the dependence of Denmark in this respect are also discussed.

International Perspectives on Social Justice in Mathematics Education, pages 191–208

FUNDAMENTAL REASONS
FOR MATHEMATICS EDUCATION

Analysis of the history of education in Iceland reveals that for long periods there was only little emphasis on mathematics education. Cultural activities were concerned with a national heritage from medieval times, preserved in manuscripts written in the vernacular, in addition to European Latin influences. The medieval heritage included some original and translated observations and knowledge of mathematical nature, which was up to the date when written in the 12th to 14th century, but became outdated with time. From 1300–1800 neither original nor recent foreign mathematical knowledge is known to have been studied in Iceland, except for land surveying and map-making in the late 16th century and 17th century.

Can explanations for this fact be supplied? The Icelanders were extremely poor, but the educational elite never lost the sight of maintaining contact with the European culture. The history paints picture of a nation living on the boarder of the habitable world, putting pride in cultural activities, but selecting what it felt it could use from the European culture and leaving out other factors. Why did it leave out mathematics for long periods of time?

In order to answer the question, the development will be measured by the fundamental reasons for mathematics education as identified by Prof. Mogens Niss:

> Analyses of mathematics education from historical and contemporary perspectives show that in essence there are just a few types of fundamental reasons for mathematics education. They include the following:
>
> - contributing to the *technological and socio-economic development* of society at large, either as such or in competition with other societies/countries;
> - contributing to *society's political, ideological and cultural maintenance and development*, again either as such or in competition with other societies/countries;
> - providing *individuals with prerequisites which may help them to cope with life* in various spheres in which they live: education or occupation; private life; social life; life as a citizen.[1]

Did Icelandic society need mathematics education for its economical or cultural development, did it cultivate mathematics for its own sake, or did individuals need to be provided with mathematical prerequisites to cope with their private or professional lives? The fundamental reasons for mathematics education or its absence in Icelandic society at each particular period of time will be analysed.

ICELANDIC SOCIETY, ORIGIN AND STRUCTURE

Iceland is an island in the Northern-Atlantic slightly larger than Ireland. It was settled by Scandinavian tribes in the Viking Age. In the 10th century, the inhabitants established their own free state which survived for over three centuries. Iceland was under Denmark from the late 14th century. It was gradually releasing its bonds from Denmark from the 1870s when its parliament was granted legislative power, subject to the King's consent. It acquired Home Rule in 1904, sovereignty in 1918 and a republic was established in 1944. Cultural relationships with Denmark lasted still longer.

Due to harsh climate and difficult living condition the population did not grow markedly until the 20th century. It is estimated to have been 50,000–70,000 in the 11th century.[2] At its first census in 1703 it was 50,358 and by 1900 it was 78,203.[3] The culture was European, marked by the introduction of Christianity in the 11th century, the evangelic Lutheran reformation in 16th century, the Humanism in the 17th century and in the 18th century the Enlightenment, whose influence lasted into the 19th century.

From the adoption of Christianity the church ran Latin schools for boys in order to educate priests. In the early 19th century there was only one such school from which priests graduated. It was also a door to further education, usually sought at the University of Copenhagen where Icelandic students had some priory of grants for supporting themselves. All public education was built on home education under the supervision of parish priests by regulation of the 1740s.

Cultural exchanges and other contacts with Europe, such as trade, depended on sailing. Sailing from Iceland to Europe was very common in the 11th and 12th century. However, when the Icelanders submitted to the rule of the Norwegian King in 1262, a part of their agreement with the King was that he would ensure that six ships sailed to Iceland each year. This indicates that the Icelanders themselves may not have had many ships at that time, presumable for lack of timber in an increasingly exploited country and colder climate than earlier.

The end of the middle ages marked a change in Iceland of an opposite kind to that which characterized most European countries. In Europe it generally meant the beginning of greatly increased trade. In the modern age, Iceland remained outside the mainstream of trade, under the trade monopoly of the Danish King, established in 1602.[4] The monopoly became a source of handsome income for the Danish crown but also a certain safety net for the Icelanders' contact with Europe.

There were several main harbours, where markets for import and export goods were located, but no towns grew up at these ports. Workers and servants of landowners were sent to the coast at fishing seasons to harness the precious export goods, and towns or villages of independent fishermen and boat-owners were not allowed to form. Trade within the country was so small that no infrastructure existed; neither roads nor bridges were needed for any major transport.[5] The lack of internal trade contributed to the persistence of an exclusively rural self-sustaining society in Iceland for centuries after towns were established in the other Nordic countries. Episcopal sees and the great estates were situated far inland, surrounded by rivers and mountains. The sees and cathedral schools established in the 11th and early 12th century remained relatively unchanged until the beginning of the 19th century. Icelandic society thus remained stagnant into the 19th century, while the complexity of trade increased in the neighbouring countries.

The 18th century was the most difficult period in Icelandic history. In 1707 a smallpox epidemic killed a large number of people, and by the middle of the century a series of cold years with pack-ice caused a famine. The situation reached its worst during the so called Haze Famine, following a massive volcanic eruption in 1783–1785, when toxic volcanic gases and ash poisoned the grass and killed the majority of all livestock, mostly from fluorosis.[6] From the end of 1783 to the end of 1786 the population decreased from 49,753 to 39,190 or just over 10,500, one fifth of the whole population. Following these calamities the Danish government gave up the trade monopoly of Iceland to become free to any citizen of the Danish crown, as the trade rendered no revenue, but also according to a plan on behalf of the government of organized urbanization of the country. By this plan, the first town began to form at Reykjavík, the present capital, in the late 18th century.[7]

GEOGRAPHY AND NAVIGATION IN THE 16TH CENTURY

Geography and navigation were two related aspects of mathematics, extremely important to the world of the sixteenth century.[8] Bishop Guðbrandur Þorláksson (1541/42–1627) was educated in mathematical subjects at the University of Copenhagen. He made land-surveying and produced a map of Iceland and thus worked on the same kind of tasks as mathematicians in the European world, contributing to the world's knowledge of Iceland's geography. His map was introduced to the European learned world in 1590 through the mediation of a Danish researcher, Andreas Sørensen Vedel. It remained a basis for Icelandic maps into the early 18th century.[9]

Bishop Guðbrandur Þorláksson was an adherent of Humanism and a proponent of education. He saw and utilized the new technology of the printing press as a prime channel for educating the people. The bishop's eagerly pursued theological activities and promotion of the Icelandic language in his publications, such as the translated Bible, may be attributed to his desire to ensure the power and influence of the Church and the independence of Icelanders. As long as they kept their own language they would retain some independence from foreign rule by Denmark. For both ends, the publication of theological works was useful and the printing press an excellent tool. Mathematical publications had no such purpose. Mathematical publications for the general public may not have been so widespread at the turn of the 16th century in Europe either, that there would be foreign models for that task, except for calendars, such as the *Calendarium*[10] published and presumably edited by the bishop. Other mathematical books might have aimed at merchants or university professors, professions not found in Iceland.

Mathematics was probably studied at the cathedral school during Bishop Guðbrandur Þorláksson's term in office. However, Latin was the main subject of the school. It was the *lingua franca* of the European world; it was the thread that kept Iceland in contact with the civilized world, and that had to be a priority.

INITIATION OF MATHEMATICS EDUCATION

In spite of hard times in the 18th century it saw a new dawn to education. The Enlightenment had considerable influence among Danes concerned with Icelandic affairs and the Icelandic elite, educated in Denmark. Among the products of the movement were a total of six arithmetic textbooks written in Icelandic in the period 1746–1841. All of them were primarily written as aids to trade, but three of them also as general textbooks in arithmetic. They witness an increase in trade, however small on European scale, but also, that the Icelanders had become aware of that they were behind the Danes and other European in mathematics education. There were no Icelandic merchants until late 19th century.

The adherents of the Enlightenment were interested in modernizing the structure of society. The episcopal sees and one of the cathedral schools were closed down and a new school and a new see were established in the growing capital Reykjavík around the turn of the 19th century. The other cathedral school had already broken down in an earthquake during the Haze Famine calamities. The idea was to establish a school in a modern style where the teachers would receive their salaries in cash, and the pupils would receive their alms in cash from the King's funds as well.

The experiment proved a disaster. The school building was poor, as was the teaching. When the old sees were abolished, an agreement was made that a certain amount of money should be allocated to cover the costs of the new see and a school for 40 pupils. The money declined in value during the Danish inflation years arising from the Napoleonic wars in the early 19th century. The see, which had previously had huge assets, went bankrupt and in 1804 the pupils of the new school in Reykjavík could not even be adequately fed. They were sent home and the school was closed down. For a year there was no learned school in the country, while it was re-established in 1805. Then the number of pupils fell to 24.[11]

A part of this modernization in the Danish realm, inspired by the Enlightenment, was legislation in 1814 on public schools in Denmark.[12] The rural structure and lack of transport made it unthinkable to adjust such legislation to Iceland at that time. Legislation of 1880 prescribed instruction for children in arithmetic within the traditional home education prescribed by regulations of the 1740s. Legislation on public schools was first passed in 1907, then for children 10–14 years of age.

Minimum requirements of mid-18th century regulations about the ability to perform the four operations in whole numbers and fractions were not fulfilled at times in the Learned School around the year 1800.[13] For several years in the early 19th century, students from the Icelandic Learned School had to be exempted from new requirements in mathematical knowledge on behalf of the University of Copenhagen due to lack of mathematics teachers. In 1819 professors at the University of Copenhagen complained about the inadequacy of the mathematics education of the Icelandic students.[14]

By a stroke of luck a scion of the Enlightenment, Björn Gunnlaugsson, one of the finest examples of the home- and self-education tradition, had studied mathematics, merely on his own with the aid of land surveyors, and without ever being accepted at the Learned School, before he entered the University of Copenhagen. Yet this education enabled him to win a golden medal for a solution of a mathematical problem in his first year at the University of Copenhagen in 1818. After five years of study, Björn Gunnlaugsson's life was devoted to raising the level of mathematical teaching of the Learned School that had refused him earlier. Another of his great feat was his land surveying yielding a geodetic map of Iceland, the basis for maps of Iceland into the 20th century. However, Gunnlaugsson was the only Icelandic mathematician throughout the 19th century and he was isolated from the developments of mathematics in Europe. The intellectual environment and interests swayed him to philosophy and the didactics and applications of mathematics, such as land-surveying.[15]

Björn Gunnlaugsson presented his goals for mathematical education in his inauguration speech at the Learned School where he emphasized the utilitarian aspects of mathematics. It was a tool to explore nature, he said, while he also argued how mathematics could train people in logical thinking, as nowhere else was truth as easy to research and easily distinguished from falsehood.[16] Thus his personal attitude was based on acquaintance of the cultural values of mathematics. He may, however, have found it wise to emphasize the practical aspects to his fellow countrymen. His magnificent work in land surveying served as a tool for seafarers and a basis for future roads, bridges and harbours, and thus contributed to the technical development of Iceland.

REGULATIONS OF 1877

Björn Gunnlaugsson was not succeeded by anyone his equal as mathematics teacher at the Learned School. Mathematics teaching at the school declined in quality after his days. At the time of his death in 1876, new regulations for Danish learned schools had divided them into a mathematics-science stream and a language-history stream. A board, established in 1875 to make proposals about Icelandic school affairs, suggested a combination of the two streams, similar to the previous structure of the school with increased emphasis on modern languages on the cost of the antique languages.[17] The number of pupils was so small at that time that a two-stream system was not feasible. Educational authorities opted for the language–history stream in 1877. The mathematics teaching was greatly reduced, while instruction in the Danish language was increased.

Letters between the Governor of Iceland and the Minister of Icelandic Affairs in Copenhagen, discovered at the National Archives of Iceland, explain why a language-history stream option was chosen in Iceland.[18] The resulting decision may presumably be attributed to lobbying on behalf of the headmaster of the Learned School, a philologist, and some teachers in favour of the language-history stream. The main reason for maintaining European standards in mathematics education on behalf of the authorities had been to enable students to pursue studies at university level. When mathematics was no longer necessary and a language-history stream without advanced mathematics requirements became an option, it was selected for the Icelandic Learned School. There were no mathematicians to promote mathematics on the basis of a personal conviction of its utilitarian and cultural values.

The three fundamental reasons adduced by M. Niss were brought up as arguments in this case, a protracted debate lasting from 1876 until 1882. The arguments for more mathematics, presented by those, who defended the proposal of a combined stream, were that mathematics education, offered after the regulations had been put into effect, was insufficient in itself and the topics that lacked would "finalize and perfect" mathematics education in the school. More mathematics would provide students with prerequisites for further mathematical studies at institutions of higher learning, such as the Polytechnic College in Copenhagen, to become engineers and thus contribute to the technological development of society. It was also argued, that mathematics had an important role as instruction in thinking for mankind, an argument referring to contribution to society's cultural maintenance and development.

Governor Finsen, a half-Icelander who grew up in Denmark, represented the opponents, who argued that the Learned-School pupils were seeking qualifications for professional examinations in theology, medicine, law or philology, and anything else would be an extremely rare exception. Those who planned to do so had to acquire the requisite skills elsewhere. The need for engineers was not yet considered relevant, and even so, it might not have been thought unnatural that an extra year in Copenhagen would be necessary for those who were inclined to become pioneers of that kind.

The dispute ended in 1882, less than a decade before the Icelandic society stood on the threshold of a new age of technical progress, far later than its neighbouring countries. There were still neither roads, bridges, harbours nor motorized ships in that large country, and conveniences such as water systems or sewages did not exist in the growing capital of Reykjavík. Visionary national leaders might have foreseen the need for technical education, a track that demanded that the mathematics–science stream option would be maintained.

The basic reasons for excluding the mathematics-natural sciences stream were of an economic nature. It was not financially possible to divide the school of about 80 pupils in six age groups into two streams. The school was already a substantial item in the country's budget, which was run at a deficit, and paid for by the Danish government. As the number of hours could not be increased, some of the teachers must have been afraid that the hours for their subjects would be cut down and hence their own share of work. They were therefore also thinking of their own personal economics.

Also politically, more people would be immediately content with reducing the workload in mathematics instead of cutting back the amount of teaching in the ancient languages, Greek and Latin, even if Latin's role as *lingua franca* had declined in importance. The classical languages were considered necessary prerequisites for the most common professional occupation, the priesthood, as well as Latin for the medical studies, in addition to their renowned qualifications in training the mind. By comparison, mathematics had no immediate application. Furthermore, evidence exists that it was taught in such a manner in the 1870s at the Reykjavík Learned School that its purpose was invisible, and its popularity among pupils minimal.[19]

Iceland was without higher mathematical education from 1877 until 1919, during an important period of progress in public education and technical innovations, such as motorization of fishing boats and organized transport. The Learned School, the only one in the country until 1930, thus became an isolated institution in society. It hardly participated at all in the country's transition from a predominantly rural structure towards a modern industrial society, a sign of its conservatism and lack of sensitivity to society's needs.

MATHEMATICS STREAM
AT THE REYKJAVIK HIGH SCHOOL

Gradually the number of Icelandic engineers grew to have become eleven in 1912. The relation between education and technical progress began to be generally recognized by the educational elite and the authorities, which were becoming increasingly domestic with the Home Rule in 1904 and sovereignty in 1918. Training Icelanders to become engineers was realized to be more economical than hiring foreigners, who only stayed for a short time, demanded higher salaries and were less knowledgeable about the circumstances than those native born. The recently established Association of Chartered Engineers in Iceland worked at presenting the need for domestic mathematical training for prospective engineers. This finally led to the establishment of a mathematics-science stream in 1919 at the Reykjavík High School in 1919, nearly half a century later than in comparable schools in Denmark. Thus the preparation of engineers was accepted at the Reykjavík High School after the need had been recognized but it was not undertaken there in order to enhance society's technical development.

The mathematician Ólafur Daníelsson acquired a doctoral degree from the University of Copenhagen in 1909. Dr. Daníelsson, who had been building up mathematics teaching at the Teacher Training College from 1908, was appointed to lead the mathematics stream of the Reykjavík High School. Dr. Daníelsson contributed strongly to the improvement of mathematical proficiency in Iceland by educating future teachers, engineers and mathematicians to be and by writing textbooks in arithmetic, geometry and algebra for the lower secondary level. His textbooks were an invaluable effort to raise the standards of Icelandic mathematical education. Together with his former teacher student, E. Bjarnason, Dr. Daníelsson built up a coherent system of arithmetic textbooks for primary and secondary level schools in Iceland. In the forewords to his textbooks, Dr. Daníelsson expressed his views that the purpose of mathematics education was to train the mind.[20] This was unique among his fellow countrymen, who considered mathematics as a tool for solving practical problems.

The reasons for establishing a mathematics stream in 1919 were utilitarian with respect to educating engineers for technical development of the country. Its implementation was entrusted to Dr. Ólafur Daníelsson, who argued for the cultural aspects of mathematics education. After his time, mathematics fell into stagnated tracks for up to a quarter of a century, until new mathematicians had acquired the status to enhance reforms.

INTERLUDE

In the 1920s, increasing interactions between the growing public primary and lower secondary education and the learned school heritage led to a relatively large number of pupils seeking attendance to the Reykjavík High School. Other educational offers remained limited, especially in Reykjavík, a rapidly expanding town, growing from 6,700 to 38,200 inhabitants in the period 1900–1940. As a result, admission to the school became restricted to 25 new pupils a year in 1928. The school and its main mathematics teacher, Dr. Daníelsson, thereby acquired a monopoly position. Even if another high school was established at that time in the northern region, the Reykjavík School in the south west was dominant. The restrictions resulted in that the growing number of primary- and lower-secondary-schools adjusted their arithmetic syllabus to the requirements of the Reykjavík School and used arithmetic textbooks by Bjarnason and Daníelsson. In a political effort to abolish the power of the Reykjavík School by separating the lower secondary grades from it by legislation in 1946, the compromise was to adopt its previous lower secondary department syllabus as a basis to its new countrywide entrance examination. This entrance examination remained in its original form into the 1960s and did not change markedly until it was abolished in 1976. Furthermore, a state textbook publishing house was established in the late 1930s in order to offer free textbooks for all children in primary schools. Bjarnason's textbooks, created in the 1920s, were chosen for arithmetic and they remained as the only standard arithmetic textbooks for the upper primary level until 1970.

The intentions of the legislator to provide university entrance to Teacher Training College graduates as a part of the 1946 educational reform were unsuccessful. Mathematical education at the Teacher Training College did not reach the level of the high school language stream. When the national entrance examination to the high school system was established in 1946, the lower secondary schools around the country had to rely on high school mathematics stream graduates to teach mathematics. Due to the entrance restrictions in 1928–1946 the number of these graduated was limited and most of them aimed at professional education. No special education for secondary level mathematics teachers was available within the country until 1951, then as a part of education of engineers at the University of Iceland. Thus the fateful decision of only implementing a language-history stream at Reykjavík Learned School in 1877 had a long lasting effect in a chronic shortage of mathematics teachers on secondary level.

The period 1920s–1960s was characterized by intellectual isolation. It was partly self-created, research focusing on literature and history of the recently independent nation, while ignoring natural sciences. It was also partly caused by the Great Depression, World War II and economic restrictions of foreign currency in the post-war period. In addition to outdated textbooks in most subjects, the educational system suffered from lack of facilities, trained mathematics teachers, and curriculum documents and it had been stretched to its limits by the mid-1960s.

MODERN MATHEMATICS IN THE 1960S

Iceland was a founding member of the OEEC. In the late 1950s theories were introduced, initiated by the OEEC, later OECD, arguing that education, especially in mathematical subjects, was central to social and economic progress. The introduction of "modern" school mathematics, stimulated by the organization, was part of a post-war awakening in science education, often associated in the West with Sputnik.

An important seminar, arranged by the OEEC, on new thinking in school mathematics was held at Royaumont, France, in November 1959. The member countries and the United States and Canada were invited to send three delegates: an outstanding mathematician, a mathematics educator or person in charge of mathematics in the Ministry of Education, and an outstanding secondary school teacher of mathematics.[21] The seminar was attended by all the invited countries except Portugal, Spain and Iceland. The Royaumont Seminar can be seen as the beginning of a common reform movement to modernize school mathematics in the world.[22]

Originally the intentions were to lay increased emphasis on applied mathematics; discrete mathematics, probability and statistics, and vectors. Influential teacher associations advocated pedagogical theories about relations between abstract algebra and logic and children's way of thinking. A quotation from the editorial of the journal *Mathematics Teaching* in April 1958 stated that "much of the psychological work of [Swiss psychologist] Piaget suggests that many of the essential notions of modern algebra (which are regarded as a university study) have to form in the pupil's mind before he is even ready to undertake the study of number ... Such topics as the algebra of sets or relations might be taught with a profit not merely ... [at upper secondary level] but lower down the school as well."[23]

At the Royaumont seminar these theories won support and its final recommendations included a syllabus introducing mathematics as a unity, and that modern algebra should be the basic and unifying item in the subject of mathematics (see Sriraman & Törner, 2007). In the teaching of all secondary school mathematics, modern symbolism should be introduced as early as possible, as it represented concepts that bring clarity and conciseness to thinking and were unifying.

In the 1960s the Minister of Education in Iceland was also a Minister of Commerce and OEEC/OECD affairs. At a meeting held in Reykjavík in 1965 the director of the Educational Investment and Planning Programme of the OECD addressed the most prominent people of the Icelandic educational system. He explained that while education traditionally had been regarded as primarily serving cultural purposes, new concepts of the role of education had recently been developed which stressed that education was as much an integrated socio-economic sector of society and the national economy as traditional sectors. Thus a new view on education was amalgamating into educational discussion in Iceland in 1965.

Influences from the Royaumont seminar reached the Icelandic mathematical community due to personal contacts with Danish participants. Small scale mathematics education reform experiments were initiated on all school levels in 1964–1966 under professional and political influences from OECD initiatives. The proponents, lead by mathematics teacher Guðmundur Arnlaugsson, were for their part influenced by Piaget's pedagogical theories. They worked on the conviction that introduction of the basic concepts of set theory and logic would enhance deeper understanding of mathematics and the new concepts would be conducive to increased clarity and exactness in thought and arithmetic.[24] Guðmundur Arnlaugsson and his collaborators had the capacity to identify the needs, to introduce and present the new ideas, and to persuade and mobilize a large group of people to participate in a reform project.

In 1965–1966, a survey made for the Ministry of Education demonstrated that the syllabus of Icelandic lower secondary level schools in mathematical subjects lagged far behind that in the Nordic countries. The survey, the mathematics reform experiments and efforts on behalf of the Minister of Education led to a general reform of the Icelandic school system, launched with hitherto unprecedented generous support. The reform was developed within a framework of a new school research department of the Ministry of Education, established on the initiative of the OECD. Its main activities became school developmental projects, such as creating curricula and learning materials, and offering in-service courses and support to teachers. With official backing and high expectations of economic progress in governmental circles, Iceland joined the "modern" mathematics reform movement in a grand manner within the frame of a general reform of the educational system.

The reform of mathematics education in Iceland thus resulted from an interaction between two actors, heavily influenced by the OECD, at the professional and political level. The initiative came in 1964 from the individual leading high school mathematics teachers who, under the influence of the

Royaumont seminar, redefined their upper secondary school's mathematics teaching, prepared the redefinition of the lower secondary school mathematics and provided consultation to the primary level. On the other hand, the Ministry of Education established school research in 1966 which developed into a large department working on school developmental projects, including compulsory school mathematics.

The reform enjoyed a massive support from the parliament, whose members complained in 1953 that every seventh *króna* went to a school system of a questionable quality compared to the venerable home education tradition.[25] The reasons were economical, expectations of economic gain from investment into education. It proved, however, to be a long term investment, providing revenue only many decades later. Industry did not demand mathematicians or physicists in great numbers for many decades. Those young people of the first generation of a new republic, who were influenced by the "need of the society" for science-educated manpower, were mainly exposed to teaching when and if they arrived back home from abroad with their scientific education in the 1970s or later. Gradually, openings for mathematicians developed in banks and insurance companies, and still later in genealogic and biopharmaceutical enterprises.

The implementation of "modern" mathematics in Iceland was at first aimed at elite pupils on secondary level, preparing for university and college studies. That part of the reform of mathematics education proceeded fairly successfully. The primary school experiment proved more controversial. Insufficient information about more than the first 2–3 year courses in a rather hastily chosen syllabus, translated from Danish, and an unexpectedly large group of teachers and pupils participating in the project, made it difficult for only few persons to organize. The new concepts were foreign to teachers and caused unrest among parents. The disturbing elements were radical ideas of implementing university conceptions of a unification of the various branches of mathematics, through logic and set theory, into primary school mathematics. That act proved to be the beginning of the end of the most orthodox "modern" mathematics reform, also in Iceland.[26]

MATHEMATICS EDUCATION
AND INDIVIDUAL'S IMPACT- A SUMMARY

We may now turn to the question: Why did Icelanders leave out mathematics for long periods? We have seen that the most common need of the general public for mathematics, trade, was minimal. However, we have also become aware of important moments when a decision was taken to practice mathematics education for the benefit of society, supported by influential individuals who knew the capacity of mathematical education and had their vision of its cultural value. There were also moments when there were no such individuals present and mathematics was left out.

The population of Iceland is small, and so was its intellectual community for most of Iceland's 1100 years, up to the mid 1970s. The mathematical community was still smaller. In the whole of the 19th century there was only one Icelandic mathematician, Björn Gunnlaugsson, whose work as teacher and land-surveyor was an admirable and unique achievement. Dr. Ólafur Daníelsson was Björn Gunnlaugsson's successor in the 20th century, being a pupil of his grandson. Dr. Ólafur Daníelsson's influence on Icelandic mathematics education through his textbooks persisted for more than six decades. After his time, his pupil, Guðmundur Arnlaugsson, became the most influential person in mathematics education in the second half of the 20th century, and together with a colleague, took the lead in school mathematics reform activities in the late 1960s. Thus there was a long-standing tradition of individual authority in the field of mathematics education.

The impact of the presence or absence of influential individuals versus official reasons for crucial transformation of mathematics education in Iceland may be summarized as follows:

- In 1590s, Bishop Guðbrandur Þorláksson, being the most powerful person in the country, which adhered to a foreign rule, made a map of Iceland based on his scientific knowledge, but also on utilitarian aspects, that a correct map would provide Icelanders with safer trade and sailing on which the contact to European culture depended.
- In 1822, when mathematician Björn Gunnlaugsson offered himself to become mathematics teacher at Bessastaðir Learned School, the official reason for his appointment and for enhancing mathematics was to ensure the pupils' prerequisites for admission to the University of Copenhagen, while Björn Gunnlaugsson brought up utilitarian arguments and cultural aspects of mathematics education.
- In 1877, mathematics was no longer required for admission to the University of Copenhagen. No mathematician existed at the Learned School to present cultural or utilitarian arguments for the subject, and the mathematics syllabus was reduced.
- When a mathematics stream of the Reykjavík High School was established in 1919 on the urge of the Association of Engineers in Iceland and mathematician Dr. Ólafur Daníelsson, the official reason was to ensure prerequisites for engineering studies, i.e. utilitarian reasons for a rapidly industrializing society. Dr. Daníelsson's arguments for mathematics education were, however, mainly cultural, presenting mathematics as the most perfect science existing.

- In the mid-1960s when "modern" mathematics was implemented as part of the revision of the Icelandic school system, the official arguments were that education would contribute substantially to economic and social progress. The leader of the activities, Guðmundur Arnlaugsson and his collaborator, had ideological arguments in mind, that the new concepts would be conducive to increased clarity and exactness.

A leadership of influential individuals was of crucial importance at points of transformation in Iceland. A redefinition of mathematical education could take place when both the official body that was to decide upon it and the persons that were to provide the pedagogical leadership had their own vision. They may not be identical but in all cases they may be classified among the fundamental reasons identified by M. Niss.

ICELAND'S RELATIONS TO DENMARK

Many historians have attempted to evaluate the impact of Iceland's relation to Denmark for close to five centuries.[27] Did the Danish monarchy exploit Iceland? In fact, the Danes had for most of the time indifferent attitude to Iceland. Furthermore, the Icelanders kept their relative cultural independence by maintaining Icelandic as official language in schools and churches, and most officials on behalf of the Church and the Danish Crown were Icelanders. It is indisputable, however, that the King's treasury yielded a considerable income from its properties, acquired at the introduction of the Reformation, and from the monopoly trade most of the period 1602–1786. By the 18th century the authorities had begun to realize that they had to nurture that resource for it to produce the desired profit, which explains various efforts toward modernization at that time. The loss of the compensations for the assets of the episcopal sees in the inflation arising from the Danish Napoleonic wars was also extremely detrimental to the economy of Iceland, the one learned school in particular. It took the school and the episcopal see several decades to recover from that blow.

It must not, however, be ignored that the Icelandic elite, the landowners and the officials, were very conservative. It was in their interest that farming was kept self-sustaining and that people were not allowed to settle in towns and by the coast. On those conditions neither social progress could be expected nor increase in population. There might be temporary increase which would be counteracted by epidemics or periods of cold climate causing famine. It was first when trade became free and fishing could be pursued as a whole-year activity on decked boats that the population began to grow and urbanization created basis for organized public education, including arithmetic. During the 19th century, trade became more prosperous than under the trade monopoly. The Icelanders gained self-esteem to begin their battle for independence and demand compensations for their losses, continuing far into the 20th century.

In their general indifference, the Danes usually adapted their activities according to the wishes of the most influential Icelanders, if it did not entail cost. The decision to adopt language-history stream at the Learned School in 1877 was a wish of the Icelandic heads of the school. Then the Danish authorities seized the opportunity to offer the pupils more Danish to learn, replacing mathematics, even if it was a general opinion that the students' knowledge of Danish was sufficient.

The first decades of sovereignty from 1918 and complete independence from 1944 were characterized by many kinds of teething problems. There was shortage of facilities, knowledge, experience and trained personnel in many spheres. In particular there was lack of mathematics teachers. Being aware of being behind other nations in many respects had caused Icelanders a general feeling of inferiority, especially for the Danes, which took a long time to overcome. However, the 20th century has been a history of continuous, although somewhat periodic, progress and economic prosperity in Iceland. The latest novelty, that it has recently also taken off abroad, is only a continuation of that development.

NOTES

1. Niss, M. (1996): 13
2. Gunnar Karlsson (2000): 45
3. Statistics Iceland, website
4. Gunnar Karlsson (2000): 127, 138–142
5. Thorsteinsson and Jónsson (1991): 136–137
6. Björnsson (2006): 208
7. Karlsson (2000): 181–182
8. Katz, (1993): 360
9. National and University Library of Iceland. Website of maps.
10. Sæmundsson (1968)
11. Ólafsdóttir (1961): 67–93
12. Guttormsson (1990)
13. Helgason, 1935; Helgason 1907–1915
14. National Archives of Iceland: Skjalasafn kirkjustjórnarráðsins SK/4 (örk 23)
15. Guðmundsson (2003)
16. Gunnlaugsson (1993): 57–66
17. *Álitsskjal nefndarinnar í skólamálinu* (1877)
18. National Archives of Iceland, Íslenska stjórnardeildin, Skjalasafn land-shöfðingja
19. Jónsson (1883): 97–135

20. Daníelsson (1920): iii–iv
21. OEEC (1961): 7
22. Gjone, G. (1983): Vol. II, 57
23. Cooper (1985): 76
24. Arnlaugsson, G., 1966
25. Alþingistíðindi (1953)
26. Gjone, (1983), Vol. 1, 53. National Archives, 1989/S-56
27. See for example Karlsson (2000); Thorsteinsson and Jónsson (1991)

REFERENCES

Álitsskjal nefndarinnar í skólamálinu (1877). Reykjavík.

Alþingistíðindi (1953). Endurskoðun skólalöggjafar D no. 9, 365–370.

Arnlaugsson, G. (1966). *Tölur og mengi*. Reykjavík, Ríkisútgáfa námsbóka.

Arnlaugsson, G. (1967). Ný viðhorf í reikningskennslu. *Menntamál* 40 (1), 40–51. Reykjavík.

Björnsson, L. (2006). *18. öldin*. In Sigurður Líndal (Ed.): *Saga Íslands* VIII. Reykjavík, HÍB.

Cooper, B. (1985). *Renegotiating Secondary School Mathematics. A Study of Curriculum Change and Stability*. London, The Falmer Press.

Ólafur Daníelsson (1920). *Um flatarmyndir*. Reykjavík, Guðmundur Gamalíelsson.

Efnahagsstofnunin (July 1965). *Nokkur efnisatriði erinda og umræðna frá fundum um menntaáætlanagerð 2. og 3. júní, 1965*. (A report from the Economics Institution: *Several Items from Presentations and Discussions in a Meeting on Educational Planning June 2nd and 3rd 1965*). Reykjavík.

Gjone, G. (1983). *"Moderne matematikk" i skolen. Internasjonale reformbestrebelser og nasjonalt læreplanarbeid*, I–VIII. Oslo.

Guðmundsson, E. H. (2003). Björn Gunnlaugsson og náttúruspekin í Njólu. *Ritmennt. Ársrit Landsbókasafns Íslands – Háskólabókasafns*. 8, 9–78. Reykjavík.

Gunnlaugsson, B. (1993). Um nytsemi mælifræðinnar. *Fréttabréf íslenzka stærðfræðafélagsins*, 5(1), 54–66. Reykjavík, Íslenska stærðfræðafélagið.

Guttormsson, L. (1990). Fræðslumál. In I. Sigurðsson (Ed.): *Upplýsingin á Íslandi. Tíu ritgerðir.* 149–182. Reykjavík, HÍB.

Helgason, Á. (1907–1915). Frásagnir um skólalíf á Íslandi um aldamót 18. og 19. aldar. 1. Skólahættir í Skálholti og í Reykjavíkurskóla hinum forna. In *Safn til sögu Íslands og íslenskra bókmennta að fornu og nýju*, IV, 74–98. Copenhagen and Reykjavík, HÍB.

Helgason, J. [bishop] (1935). *Meistari Hálfdan. Æfi- og aldarfarslýsing frá 18. öld*. Reykjavík, E.P. Briem.

Jónsson, F. (1883). Um hinn lærða skóla á Íslandi. *Andvari* 9, 97–135. Reykjavík, Þjóðvinafélagið.

Karlsson, G. (2000). *Iceland's 1100 Years. The History of a Marginal Society*. Reykjavík, Mál og menning.

Katz, V. J. (1993). *A History of Mathematics. An Introduction*. New York, HarperCollins.

Kjartansson, H. S. (1996). History and Culture. In J. Nordal and V. Kristinsson (Ed.): *Iceland. The Republic*, 61–106. Reykjavík, Seðlabanki Íslands.

National and University Library of Iceland. Website of maps: http://kort.bok.hi.is/, accessed November 7, 2006.

National Archives of Iceland:

Íslenska stjórnardeildin. S. VI, 5. Isl. Journal 15, no. 680 *Skólamál.*

Skjalasafn landshöfðingja, LhJ 1877, N no. 621. *Tillögur ráðgjafans um reglugjörð fyrir hinn lærða skóla.*

Skjalasafn Fræðslumálaskrifstofunnar 1989/S-56. *Skólarannsóknir.*

Niss, M. (1996). Goals of Mathematics Teaching. In *International Handbook of Mathematics Education.* Part I, 11–47. Dordrecht/Boston/London: Kluwer Academic Publishers.

OEEC (1961). *New Thinking in School Mathematics,* 2nd ed. Paris.

Ólafsdóttir, N. (1961). *Baldvin Einarsson og þjóðmálastarf hans.* Reykjavík, HÍB.

Sriraman,B., & Törner, G. (2007). Political Union/ Mathematics Education Disunion: Building Bridges in European Didactic Traditions. (In press) in L. English (Editor). *Handbook of International Research in Mathematics Education (2nd Edition).* Mahwah, NJ: Erlbaum.

Statistics Iceland, website: www.hagstofan.is, accessed November 7 2006.

Sæmundsson, Th. (Ed.) (1968). *Calendarium* (First printed in 1597). Reykjavík, Bókaútgáfa Menningarsjóðs og Þjóðvinafélagsins.

Thorsteinsson, B., and Jónsson, B. (1991). *Íslandssaga til okkar daga.* Reykjavík, Sögufélag.

CHAPTER 13

"BEFORE YOU DIVIDE, YOU HAVE TO ADD"

Inter-Viewing Indian Students' Foregrounds

Ole Skovsmose, Helle Alrø and Paola Valero
in collaboration with
Ana Paula Silvério and Pedro Paulo Scandiuzzi

ABSTRACT

Students' cultural diversity is an important factor to consider in a mathematics education concerned with equity. We argue that the significance of mathematics education is not only given by the understanding of mathematical concepts but also by the students' foreground, that is, the students' perception of their future possibilities in life as made apparent to the individual by his/her socio-political context. For students in a cultural borderline position different reasons and intentions for engaging in mathematics learning may be related to the construction of meaning in mathematics. Through inter-viewing Brazilian Indian students' foregrounds, we illuminate the different types of significance given to mathematics education in their particular situation.

International Perspectives on Social Justice in Mathematics Education, pages 209–230

INTRODUCTION

That mathematics education should pay attention to students' cultural diversity is not any new premise for a practice concerned with equity. During the last decade many different research and development initiatives have provided insight into how to conceptualise and realise in practice sensitivity for this issue. It has been evident that the initial focus on mathematics as a cultural activity (e.g. Bishop, 1988) with its emphasis on how different human groups develop mathematical notions has been enlarged to include a wider perception of the different actors who play a role in the practices of teaching and learning (e.g., Abreu, Bishop & Presmeg (Ed.), 2002). The ethnomathematics program has also contributed in an understanding of how different human groups generate and interact with mathematics.

One of the focuses of research concerned with cultural diversity is addressing processes of exclusion associated with traditional mathematics teaching and learning in relation to certain groups of students. Exclusion and inequalities in mathematics classrooms operate on the grounds of students' social class, gender, ability, language, ethnicity and culture. We are particularly interested in issues of culture and ethnicity since this factor has a growing impact on students' exclusion from participation in mathematics learning, at a time of growing heterogeneity among students in classrooms. As internationalization and globalization advance, diversity of people in local communities increase, and so does the risk of reproducing social patterns of exclusion in mathematics classrooms.

In many societies cultural and ethnical diversity has increased with migration of peoples. Normally immigration and emigration are discussed with respect to the moving of groups of people into different geographical spaces from their native ones. The notions of immigration and emigration signify the perspective from which we are looking at the situation. When we see people as entering our society, we talk about immigrants, and when we see people as leaving our society, we talk about emigrants. The situation, however, could be 'inverse', when the actual immigration (or emigration) is not caused by the moving of the group of people in question, but because of changes in the whole socio-political and economic environment. For many indigenous peoples in many countries in the world, their culture and environment has been overrun by external forces. This is not an exception for many indigenous communities in Brazil.

During the time of colonisation the invading powers tried to make slaves out of indigenous people, but Indians were difficult to enslave. They knew the environment all too well, and could escape slavery by withdrawing deeper and deeper into the forest, leaving behind a huge land for the invading people to take over. This withdrawal of indigenous peoples seems to have continued ever since, although compensated by the Brazilian government turning some areas into a *reserva indigena* (Indian reserve). Here the indigenous people may experience a *borderland position*.[1] On the one hand, they can preserve some of their traditions and ways of living, although they can do so only in an environment, which seems always in danger of being overrun by industrial interest: mining, the exploitation of the forest, or farming, all of which always try to carve deeper into the reserva indigena. On the other hand, the Indians are well aware of the strengths and powers of the Western civilisation, for instance in terms of possibilities for improving life conditions in general, and health care in particular.[2] In the case of indigenous peoples in Brazil, it is the rest of the world that, so to say, is moving by turning up right outside their natural environment. The consequence, however, is the same: a group of people experiences a borderland position with references to two different cultures.

How does experiencing a borderland position influence students' motives for learning? We expect that in general motives for learning are related to a person's background as well as his or her foreground. Background refers to a persons or a group of persons' cultural and socio-political roots; and foreground refers to a person's interpretation of learning and 'life' opportunities, which the socio-political context seems to make available. For a student in a borderland position, however, background and foreground as well as the relationship between them could easily be structured by conflicting priorities and possibilities. What does the socio-political context allow students in a borderland position to hope for and to expect as being part of their ('realistic') possibilities?

In this sense indigenous students' situation is similar to the situation of (other) immigrant students. Thus, we might be able to learn more about the situation of immigrant students by considering more carefully the case of Brazilian indigenous students. In this country there has been a strong concern for formulating what a mathematics education facing cultural complexity and diversity could mean. Research literature, not least the one developed around the ethnomathematical programme, has demonstrated a great sensitivity with respect to cultural diversities. It has been emphasised that education cannot remain a form of cultural invasion; rather, an activity where cultural diversity is respected and taken seriously into account.

In this context, the issue of *meaning* or *significance* becomes important. We see meaning and significance of mathematical learning activities as related to the students' foreground and background. Therefore, we find that it is of paramount importance to investigate students' foregrounds in order clarify motives for learning. To students in a borderland position these motives for learning might reveal a deeper complexity and they might include conflicts and dilemmas, which we hope to be able to clarify further. This is our intention in this chapter.

ETHNOMATHEMATICS AND MEANING

The ethnomathematical programme, introduced world wide by Ubiratan D'Ambrosio's plenary at ICME-5 in Adelaide, puts into focus the idea that mathematics operates in a variety of cultural settings. This programme broadens the concept of mathematics: not only can we experience mathematics in textbooks and in mathematical research journals, but also in any form of handicrafts, for instance as represented in construction of houses or of boats for sailing at the Amazon River. Mathematics can be integrated in tools, craftwork, arts, routines. It can be part of a chair as well as of a computer. D'Ambrosio has interpreted the notion of ethnomathematics by considering its three conceptual elements: ethno-mathema-tics. 'Ethno' refers to people; 'mathema' to understanding; while 'tics' refers to techniques as well as to art.[3] Thus, ethno-mathema-tics refers to culturally embedded ways of understanding. It must be noted that the notion of 'mathema' is broader than 'mathematics' as we normally consider it; that 'ethno' has to be understood as people/culture, and that it does not include any reference to 'ethnicity' (understood as a racial category).[4]

According to the conceptual delineation of ethnomathematics, we could talk about the mathematics of bakers, carpenters, street children, vendors, bank assistants; we could talk about the mathematics of the Incas, as well as of tele-engineers, system developers, dentists, statisticians and mathematicians; and we could also talk about the mathematics of students in a borderland position. In other words, we adopt the idea that every community develops a particular mathematical practice; and such a practice is meaningful for participants in it. We are not interested in digging out the characteristics of the mathematics of indigenous students. We are rather willing to investigate the perception that they have about who they are, their life in the reserva indigena, schooling and, in particular, the meaning they give to learning mathematics. In this perception we hope to find the motives that they have for learning mathematics. As mentioned before, such motives could be found in the background as well as in the foreground of the students.

'Foreground' refers to a person's interpretation of learning and 'life' opportunities, which the socio-political context seems to make available.[5] Thus, the foreground is not any a priori given to the person; it is a personal interpreted experience of possibilities. We talk about 'multiple foregrounds', as a foreground can be acted out in different ways, depending on the situation. A person does not necessarily maintain a universal foreground, but he or she could switch between different foregrounds. To a teenager dreamy and realistic elements may be mixing. Depending on the situation different foregrounds could be brought in operation, and in this way serve as motives for actions and for bringing intentions-in-learning.

Foregrounds are changing, and we can observe a strong discontinuity. Suddenly, a new way of looking at one's possibilities may emerge. This can, for instance, be due to change in the social environment. New motives for learning can emerge, apparently out of the nowhere. A *foreground is not a particular 'thing'*, which we as researchers could hope to discover in a proper way. It does not make sense to ask: What is the real foreground of a person? A foreground is a dynamic interpretation of person's or a group of persons' future possibilities.

Apparently a background is a more stable unit than a foreground. However, also a background is an interpreted phenomenon. As one can see one's possibilities in different ways, so can also one's background be interpreted and reinterpreted. Sometimes it can appear a valuable resource; sometimes it appears to be an obstacle for getting on in life. Both foregrounds and backgrounds are resources for people to construct motives for learning. From these resources intentions can be put into learning.

For students in a borderland position one could expect that conflicting elements in and between foregrounds and backgrounds would appear in forms of approaching learning. These elements might influence the way students see meaning in education. In what follows we inter-view some students in an Indian village in order to illuminate this point.

THE VILLAGE KOPENOTY

Kopenoty is located in a reserva indigena, in the centre of the State of São Paulo, about 30 kilometres from Baurú, a city of about 500.000 inhabitants. Baurú has several universities, a department of the State University of São Paulo being one of them. Further there are many Faculties in Baurú one being Faculdade do Sagrado Coração. In the State of São Paulo there are several other reservas indigenas, most of them close to the coast line.

Kopenoty has a school built by the government of the state. This is a simple brick building, although a huge improvement compared to the round straw roof, which could provide some shadow for the few benches that up to then had made up the school facilities. The houses in the village are very small. They are hiding around in the landscape. Recently electricity has been installed. In the middle of the village, we see a small soccer field.

It is difficult for an outsider to get access to this village. The Indians are suspicious of any white person trying to get access. They could be suspicious of the white person's motives. They could also simply be tired of having interested people sneaking around. In this case access was granted by the chief of the village. And he, in turn, would consult the federal department responsible for the security of the reserva indigena. Only after such procedures one could get the permission to enter.

INTER-VIEWING FOREGROUNDS AT KOPENOTY

When investigating foregrounds we consider the relationship between the interviewer and the interviewee. We suppose that foregrounds exist as constructions that cannot necessarily be found in any 'true' or 'pure' form. Therefore, it is legitimate for the interviewer to engage in an active interviewing as a way of revealing and co-constructing multiple foregrounds. Steiner Kvale (1996) has used the expression inter-viewing.[6] We find that this elegant formulation of 'seeing together' condenses nicely our approach to researching foregrounds. From the part of the researcher, there is no hidden agenda, i.e. something in the research design that we keep secret from the person whose foreground is investigated in order to obtain 'validity' in the research. This makes it possible to consider *dialogue* as an adequate research approach (see for example Stentoft, 2005). Through dialogue and collaboration perspectives can be stated, examined and challenged, and the participants can get to examine their own thinking more clearly. Therefore, we think of dialogue as a methodology for inter-viewing foregrounds.

The inter-view with the students was conducted by Ana Paula Silvério, who was granted the access to the reserva indigena. One reason is that Ana Paula had a good contact with the group since she has worked with teacher education there. Pedro Paulo Scandiuzzi has provided further information about the people from Kopenoty. He has worked for many years with indigenous people in order to develop a mathematics education with references to their cultural environment. Some time after the inter-view was conducted Ana Paula and Ole visited the village. However at that time the students were not available for any follow up inter-viewing. So what we are presenting here we have been seeing through the eyes of Ana Paula, Pedro Paulo and Ole, and we will come to listen to the students' voices through the inter-views conducted by Ana Paula and translated from Portuguese by Anne Kepple.

The day and the scene of the inter-view was described by Ana Paula in the following way:

> The inter-view in the Kopenoty village was scheduled for the 26th of September, 2004, at 9:00am on a Sunday. The night before, the Indigenous people had participated in a party sponsored by a candidate for city council, with a lot of food, drink, and forró (dance music from north-east Brazil). I waited until the scheduled time, but by 10:00, no one had arrived for the inter-view. While I waited, I talked with Mauria (white woman married to an Indian named Chicão who works for FUNAI [the Federal Bureau of Indian Affairs]. She suggested we go to the residences of some of the young people to conduct the work, which we did, not having any other option. The inter-view didn't have the expected result, as when we arrived at their houses, despite

the good reception, they had to stop what they were doing to talk to us. I felt that they were very intimidated, and this made it difficult for me to do what I had planned to do as the inter-viewer: conduct an informal, relaxed inter-view. I also think that Mauria's presence made the young people even more introverted, in addition to the tape recorder used to record the conversation, which compromised the ease/agility of the interviewees. Initially, the idea was to conduct an inter-view with two couples, which was not possible. I was able to inter-view only two men and one woman. I don't know if Mauria's comments should be taken into consideration, as with each answer given by the Indians, she interfered.

In this description of the context of the inter-view Ana Paula cannot hide her disappointment about the whole situation. She seems to have expected something different and she makes her reservations. Anyway, in what follows we shall have a close look at the inter-viewing.

The Students

Ana Paula first asked the students to talk about their city and neighbour-hood. She told that they could try to describe things for a person who did not know their village:[7]

Ana Paula:	Where do you live? Talk about your city, neighbours. Imagine that you have to tell this to a person who lives far away from here—in Denmark, let's say.
Maria Luiza:	I'm Maria Luiza. I'm 17 years old. I've lived here in the village Kopenoty since I was born. It's in the municipality of Avai, and is close to Baurú. The village is quite large, and we are all family here. I live with my father and my sister. Most of my friends are from the village here, and I also have friends from Baurú and other schools.
Patrick:	I'm 17 years old. I was born in the city and used to come here on the weekends. When I was a bit older, I decided to come live here in the village with my grandfather. I couldn't get accustomed to the ways of the city; everything here in the village is calmer. The work is harder, because we have to work in the field. The village is big, and there is plenty to do here. There's a reservoir where we swim, and a soccer field. I'd never leave the village now. My girlfriend is pregnant, and we're going to live together. We plant

and harvest manioc here. We have other things, too, like the vegetable garden.

Matheus: My name is Matheus. I'm 16 years old. I've lived here in Kopenoty since I was born. My mother used to live in another village, but when she and my father got married, she came to live in his house. The other village is Nimuendajú. I always go there; I have relatives there, too. But my friends are from right here. We play soccer every Sunday in the soccer field. We play against some teams from Avaí. We have lots of parties here, too, and at the parties, we dance forró.

The students all emphasize that they are strongly located in the village. Although Patrick is not born in the village he has no doubt that he will stay there: "I'd never leave the village now." He "couldn't get accustomed to the ways of the city," he says. He probably refers to pace and noise, as he declares that "the village is calmer." The others have lived in the village since they were born and have all their relatives, friends, and activities there. That Kopenoty is positioned in a borderland position is underlined by the fact that Maria Luiza has many friends in the village as well as in Baurú.

In the following we will focus at the students' foregrounds. We will make references to the school, their friends, how they experiences mathematics, how they see their future, and how they see mathematics with respect to this future. We will discuss these elements as part of their landscape of learning mathematics.[8]

The School

Ana Paula: What do you think about going to school? What do you like and what do you not like about school?

Maria Luiza: I like to go to school. There are a lot of different kinds of people there. I just don't like going to Baurú to study. It's far away, and you get too tired. Since it's at night, I get very sleepy.

Patrick: I agree that it's very tiring to leave the village here and go study in Baurú. Maybe if the classes were in the school here [in the village] we would get more out of it. I get tired and I have to force myself to pay attention to the teachers. But since I like to study, I go.

Matheus: I think, too, that it's tiring to leave the village to go to Baurú or Avaí to study. If it were here in the village, I think even people who have dropped out would go

back to school. I have lots of friends who don't like to
go to the city to study.

The students agree that they do not like to go to Baurú in order to go to
school. They are about to finish the upper secondary school, which makes
choices about the future important: What to think of their possible further
studies? Continuing in Baurú signifies entering universities or faculties, and
thereby entering a radical different environment. Remaining in the village
means remaining in the indigenous environment. The choice is existential:
two radical different life-lines could be formed. The remoteness of Baurú is
expressed in terms of distance, but this distance can be understood first of
all as a cultural distance. The students share the wish of taking an education
in their own village instead of having to join another culture in Baurú. This
seems to bother them. And why do they have to study at night? We do not
get the answer to this question, but there might be a hidden conflict here.

Friends

Answering the previous questions all three students touch upon the issue
of friends. Friends seem to be an important element in their thoughts and
priorities.

Ana Paula: Who are your friends? What do you like to do with
your friends? Do you talk about the future sometimes? What do you talk about?

Maria Luiza: My friends are Fabiana, who is my cousin, and Eluza.
We stay in the village more than go to the city. We
go to the parties and dance forró. When I go to Fabiana's house, we like to listen to the radio. Now, I
don't remember exactly what we talk about the future. But I know Eluza wants to be a dentist. I want to
be a teacher and teach little kids. It's because I like
children, that's why.

Patrick: My real friend is my grandfather. We are very good
friends. Since I go to work in the field with him, we
talk the whole day long. He tells me that he would like
me to get an education. He thinks that I could help
our people more if I study. But I have other friends,
too. I have one from the college prep course who is
cool. His name is Marcos and he wants to study how
to work with computers. I want to be a nurse.

> **Matheus:** I have lots of friends—those from my class, and others from the village. We make plans with our friends from the city to go out in the square there in Avaí. Sometimes they come to the forró here in the village, too. Talk about the future? Sometimes we talk, yes, but I still don't know exactly what I want. In the third year of high school I'll have to decide, so later on I'll resolve that.

Maria Luiza mentions her two best friends by name. They prefer to stay in the village and join the local activities: parties, they dance forró (a popular Brazilian dance, and not any indigenous dance) and listen to the radio. Maria Luiza knows that one of her friends wants to become a dentist, and she, herself, wants to become a teacher.

Patrick's best friend is his grandfather. He previously mentioned that he came to live in the village with his grandfather. And now he mentions that his grandfather want him to have an education in order to "help our people more." The two of them obviously have a strong relationship. Patrick and his grandfather work together in the fields and they "talk the whole day long." Patrick has also got other friends. He knows one who is working with computers, and he himself wants to become a nurse.

Thus, both Maria Luiza and Patrick have chosen helping professions for their future life. And they want to use their education in the village for the indigenous people.

Matheus has many friends, both in the class and in the village. They are meeting here and there, but time has not come for Matheus to consider his future life. Schooling brings young people to a crossroad. In a country like Brazil we find enormous differences between salaries. Different career opportunities really signify radical different life opportunities. For the indigenous students the differences are even more dramatic. The crossroad represents two radical different opportunities in life: not only in terms of economy but also in terms of cultural choices.

Mathematics

> **Ana Paula:** What are you doing in mathematics class? What have you already learned in mathematics?
>
> **Maria Luiza:** We're learning [algebraic] equations. I don't like this subject very much. I think it's difficult—I can't get it into my head. I liked doing calculations, but I didn't know it very well. We had sets, theorems, natural numbers, too, and that delta, which is very difficult.

Patrick: In the college prep course, I learn all the subjects taught in high school. I learned the theorems, [algebraic] equations, roots, how to transform meters into kilometers. I learn a little of everything, since they are the subjects you need for the college entrance exams. When I was young, I liked to do problems and multiplication. I have trigonometry, which is difficult, and is often on the test.

Matheus: I think it's kind of boring, this subject. I think it's because I don't pay very much attention. I don't like the teacher very much, either. She goes too fast; there is hardly enough time. I am learning to find the area of a square or a rectangle, but I think that's geometry. I am also learning cathetus and hypotenuse. The teacher is giving it to us in school. But it's very difficult for me to learn.

Maria Luiza is learning about equations, and she does not like this. It is difficult, and such topics "does not enter her head." She likes to do calculations, but they have also set theory, theorems, natural numbers, and a formula which contains the Greek letter "Δ."[9] Such things do not make much sense to her.

Patrick experiences things quite differently. He seems to have grasped most of the things, and he liked mathematics, also when he was younger. The main motivation, however, seems to be the 'vestibular', the examination for entering universities, and he explicitly refers to the topics they are addressing in mathematics as being of relevance for the vestibular, which determines the entrance to a university or not. In Brazil the upper secondary school does not conclude with a formal examination. Instead each and every university organise its own entry test. The students, then, have to sign in for the tests at the particular university, and hope that the result of the test is good enough for being accepted.[10]

Matheus thinks that mathematics is boring. He does neither like the content nor the teacher, who does everything too fast, as if there was no time for anything. Matheus learns about areas of squares and of rectangles, and he remarks that this must be geometry, indicating that geometry might be different from mathematics. He refers to cathetus and hypotenuse, so we could guess that the Pythagorean Theorem has passed the blackboard. It all appears to be rather difficult to Matheus.

In these answers it is difficult to hear any comments that reflect that we have to do with a teaching of mathematics in a particular context. Initial comments about mathematics seem universal. It appears that mathematics, as taught according to the school mathematics tradition, is as remote from the students' reality, regardless if this reality is to be found among Indian students in a village in Brazil or among Danish students in Copenhagen. The situation, however, might be different if we leave the school mathematics tradition,[11] but for the moment we are in no position to make any observations about this situation. It simply appears that the school mathematics tradition operates world-wide, and that it has reached deep into the Indian *reserva indigena* with the same effects for the students here as for students anywhere else.

However, when we try to see mathematics outside the classroom, then the different contexts could make a difference. For although, somehow, the mathematics classroom structured by the school mathematics tradition might look almost the same around the world, then the socio-political and cultural contexts in which the students are situated are different. So what to expect of answers to the following question:

> **Ana Paula:** Outside of school, do you ever use numbers, do calculations, count, make estimations? If so, what kinds of calculations do you do? In what situations is it necessary?[12]
>
> **Maria Luiza:** In school with the teacher, we use it a lot. For example, we use division to divide the materials among the children, since if one has more than another, you see how they fight, right? We also use quantity a lot, to know how much to make for lunch and snacks. In arts and crafts we also add and divide.
>
> **Patrick:** In the fields, we use division a lot, too. We divide the land according to the seed we are going to plant. We also divide our profits and expenses. We use division for everything, and addition, too, since before you divide, you have to add.
>
> **Matheus:** I agree with Patrick. In the fields, or there in the vegetable garden, we use division to divide the area to be planted and the seeds.

Maria Luiza first remarks that one uses numbers in school, but her next utterance shows that she has not the mathematics lessons in mind. She refers to mathematical operation of division as related to the process of fair sharing among children. She adds that calculations are used for the division of snacks and sandwiches and that workers use division much. Patrick continues that people in farming are doing divisions, for instance when one have to divide a field according to what one is going to plant. People are also dividing the harvest, and here one needs addition as well as people have to add up everything before it is divided. Matheus agrees with Patrick. Division is the most common operation in every-day life. To what extent this division is experienced to have much in common with the division in mathematics education is not illuminated, although this might in fact be the case.

The Future

Certainly, the students see some mathematics (in the terms of division) in their daily-life environment. But could they see mathematics in the perspective of their future? The idea, which might not be brought clearly through in this formulation is that they need not be restricted in their imagination. They could allow themselves to be rather dreamy and with great hope.

Ana Paula: What do you want to be/do in the future? Where would you like to be living? You can say whatever comes to your mind.

Maria Luiza: I want to be a teacher in the school in the village; give classes to the children and, who knows later on, be a director. But first I have to go to college [the university]. I want to stay always here in the village with my family and my friends.

Patrick: I want to be a nurse to help all my Indian relatives to be more healthy. What I really want is to work in the health post of FUNAI. I think I can even manage to achieve it. But first I have to study for that. I don't want to go live in the city.

Matheus: I don't think I want to live in some other place, no. I like the village. I want to continue working in agriculture.

Maria Luiza repeats her wish to become a teacher and to stay in the village among her friends. Patrick repeats his wish to become a nurse in order to help his whole family to get a better health. Furthermore, he wants to work in the health post of FUNAI. He thinks that he is able to manage, but, as he emphasises, first he must study in school, and this is going to be outside the village for a while. Matheus agrees about the wish to stay in the village, and he wants to continue to work with farming.[13]

Then Ana Paula emphasises that they can be realistic in the expectations:

Ana Paula: Talking realistically, what do you think the future will be like?

Maria Luiza: I don't really know, but I would like it if everyone had the same rights. I want everyone to be equal, even with cultural differences.

> **Patrick:** What Maria Luiza said is true. It would be good if everyone was equal, and everyone respected the other, as we all have different ways of thinking. I would like it if the children of the village didn't have to suffer prejudice because of their race; it may not seem like it, but there is a lot of that here. Just that fact that the children aren't interested in learning the native crafts and mother tongue—they are denying their parents' culture.
>
> **Matheus:** This is true. I have friends who only want to date white girls in the city. This is prejudice, too. I want that to end in the future.

The original intention was that the students now had to consider their situation and their expectations in more realistic terms. However, Maria Luiza interprets Ana Paula's question differently. Maybe because they already have been realistic in their previous answers. She makes a very intense statement about what she would hope for of the future: "I would like it if everyone had the same rights. I want everyone to be equal, even with cultural differences." This indicates that she does not experience equality now, and that she is well aware of the problems of cultural differences.

Patrick agrees:

Everybody should be equal, and we should respect each other. In particular, he does not want that any children should feel inferior because of their race. And he adds that when children do not have interest in learning the (local) handicrafts and their mother tongue, they are negating their own culture. Matheus agrees and refers to one of his friends who only wants to date a girl from the city. This is, Matheus emphasises, prejudice.

In these statements, we see an example of how the problems of racism are experienced by the young Indians of Brazil. This racism could take the form of self-negation or of lost self-esteem. What is outside the local village, being a possible girl friend, could get paramount values. The Portuguese language, being the language of power, could be the most appreciated language. One counter-move could to be a re-establishment of self-esteem. And here we see a crucial element in the ethnomathematical approach, which, in particular, has emphasised the essence of establishing a balanced view of the different forms of knowledge: between curriculum-based school knowledge and the cultural-based knowledge.

What is touched upon here has not only to do with students in an Indian community in Brazil. It has to do with any group of students who come to operate and to learn in a borderland position. There could easily exist an imbalance between the different cultural settings, which are referents for the borderland position. And this imbalance could, for instance, provoke a low self-esteem. The general point is that motives for learning can be facilitated as well as obstructed by a variety of foreground and background factors experienced by students in a borderland position.

Mathematics and the Future

Ana Paula:	Do you see any relevance for mathematics (or knowing how to count, make estimates, relate quantities, etc.) for your future?
Maria Luiza:	I think so. For everything in life, we're going to use mathematics. To deal with money even, we have to have a notion of values.
Patrick:	Yes. Everything we learn in school, be it mathematics, Portuguese, or biology. We use the basic notions of each one. Some things, I think, have nothing to do with anything, like the more difficult subjects.
Matheus:	You saw yourself in the field that it's necessary to count the seedlings, divide up the seeds and the land. And since I want to continue working in agriculture, that means that I'll always be using mathematics.

First, we should notice that they make a re-interpretation of future. Now it is not any longer the grander hopes concerning greater equality and mutual respect, which defines 'future'. Instead they now take the more limited perspective, assumed in the format of the inter-view. Maria Luiza emphasises that they are going to use mathematics for "everything in life," and she refers to the issue of dealing with money. Patrick agrees, although he finds that some of the more difficult things in mathematics useless. Matheus refers to agriculture, where he is always going to use mathematics.

Ana Paula:	Do you see any connection between the mathematics you are studying in school and what you would like to do in the future?
Maria Luiza:	Ah, very little. For example, what use is an [algebraic] equation if I'm going to be a primary school teacher? I'm not going to teach this to my students. But I'm going to teach division, multiplication, addition, and subtraction.
Patrick:	And really, what good is trigonometry, theorems, and roots if it doesn't get me into the university? But in nursing itself, I don't think so. I'll need to know

plenty about medicines, schedules, and for that I'll use division; to ask the patient to take a given medicine every 12 hours, it's necessary to know how much medicine for the weight of each patient.

Matheus: I don't think that the cathetus and the hypotenuse are very useful for the farmer. Maybe for me to pass the college entrance exams I should know it, but on the farm, you don't use any of that.

This seems to make a difference. Thus, Maria Luiza thinks that there would be very little use of the mathematics she is learning in school with respect to her future. She is certainly not going to teach her students equations, but calculations. Patrick thinks that entering the Faculty he would use the mathematics he is learning for the moment. Clearly enough he sees that the topics are relevant from the perspective of being able to pass the vestibular. But he adds that for the job of being a nurse, he could not come to think of any use. The relevance of numbers, he sees, has to do with measuring out the correct quantity of medicine, organising the time schedule for taking medicine, and so on. But things like trigonometry would not be of much use. Matheus also refers to the use of mathematics in order to pass the vestibular. But in the fields, he knows about, there no use of Pythagoras.

THE SIGNIFICANCE OF MATHEMATICS
FROM A BORDERLAND POSITION

It is interesting to observe how the *significance of mathematics* is expressed through the inter-view. In the beginning it appears that mathematics, as presented in accordance with the school mathematics tradition, is without much deep felt significance. However, the students seem ready to ascribe different forms of significance to mathematics.

First, we can talk about the *instrumental significance* of mathematics. In fact most of what is taught in school, according to Patrick, is relevant when we consider it from the perspective of the vestibular: all kinds of questions could be included in the vestibular. So, if one wants to get further on in life, there is no doubt that mathematics has a big instrumental significance.[14] The instrumental significance appears to be a life condition for students.

One could then consider if instrumental significance is different for different groups of students. If the results of mathematical tests, based on the particular knowledge exercised within the school mathematics tradition, mean a difference for the future of the students, then one could talk about different forms of instrumental significance. Thus, the instrumental significance for, say, children belonging to a dominant cultural group might be different from the instrumental significance given by a group of immigrant children, when we consider a particular teaching-learning content. For Indian students in Brazil, further education provides the main route out of the village. And if education should in fact come to work in this respect, then it is essential to come to master the knowledge, which becomes the code for moving further in the educational system. In particular, to students in a borderland position, instrumental significance can be of paramount importance. However, this is certainly not the purpose of the inter-viewed students in this paper. The instrumental significance is related to passing exams, but they do not want to leave the village. They want an education in order to be able to help their people.

We could consider other forms of significance. The content of mathematics could also appear significant when it is related to out-of-school practices. A practice that everybody, and every student in particular, is familiar with could establish *daily-life significance*.[15] During the inter-view, the students made statements about this form of significance. Thus the students have no difficulties in relating the processes of division to daily life practices, for instance with respect to doing work in the fields. In such cases the students seem to recognize the daily life significance of mathematics. However this daily-life significance applies only to a very restricted portion of what the students learn mathematics in school. They emphasise that a great majority of topics seem without any daily-life significance.

With respect to daily-life significance, we have observed many possibilities for cultural dominance. For instance, by ignoring that certain forms of daily-life significance are relevant to address in school, while other forms appears irrelevant.[16] The strength of the ethnomathematical position is that the daily-life significance of the mathematical activities presented for students is carefully dealt with showing a great respect for cultural values. But the way daily-life significance is dealt with in the cultural priorities of school concerns could also include a cultural dominance and affect the students' self-esteem.

During the inter-view, the students shortly touch upon what we could call *expected work-practice significance*. Here we refer to the work practices, which the students might want to enter. So what could be the significance of the issues learned in school, when we consider the work practice of a nurse or of a person working in agriculture? This appears to be cloudy to the students. It is not negated that there could be such forms of significance, hiding here and there in the curriculum. But nothing gets clearly through. One could also talk about *work-practice significance* with the "expected." Thus, one could be aware of a significance of a mathematical insight with respect to bridge construction, estimations of degrees of pollution, cryptography, without assuming that such work-practices are "expected."

There could be other forms of significance. The students made strong statements about the important of equality. They wanted society to become more equal. To what extent, if any, a mathematics education could provide any input to such a development was not addressed. The question to what extent one could imagine that a mathematics education, maybe positioned in great distance of the school mathematics tradition, could provide any input to a general improvement of society was not formulated. One could talk about a *socio-political significance of* mathematics as well as about a *significance for* a *critical citizenship*.[17] But such forms of significance were not touched explicitly upon during the inter-view. It must, however, be noticed that the statement about the relevance of further education for helping the people in the village includes a strong statement about solidarity. There might not be a long distance before we could reach a significance for critical citizenship as a possible experienced significance for students.

When students try to see the meaning of what they are doing in mathematics this could be done in different ways. And we have touched upon instrumental, daily-life life, expected work-practice, socio-political and historical significance as well as significance for critical citizenship. These types of significance are interrelated. Naturally, we could expect other forms of significance to emerge from the background as well as from the foreground of the students. We must consider the situation with respect to different groups of students. For instance, what might be a daily-life significance or an instrumental significance for a group of students might depend on the context of the students. And one such particular context is experienced by students in a borderland position, being, for instance, Indian students in Brazil or immigrant students in, say, Denmark.

The significance related to a critical citizenship might also vary from context to context. In fact one could interpret the formulation: "before you divide, you have to add" as a strong expression of solidarity. We have to add (whatever we have) in order to divide (equally) what we have brought together.

NOTES

1. The "borderline" metaphor has been used in research dealing with cultural diversity. We see *borderline* as a space of individual and social exchange where the meaning of difference is negotiated. A *borderline position* is therefore a relational situation where individuals meet their social environment and come to terms with the multiple choices that cultural diversity makes available for them. For further discussion of related notions see (Chang, 1999; MacDonald & Bernardo, 2005).
2. We do not consider the exceptional cases where Indian groups live in a much more profound isolation.

3. See, for instance, D'Ambrosio (2001). See Ribeiro, Domite & Ferriera (Eds.) (2004) for a recent contribution to the ethnomathematical research programme. See also Gerdes (1996); Powell & Frankenstein (Eds.) (1997); and Knijnik (1998, 2002a, 2002b, 2002c). It should be noticed as well that 'tics', in the interpretation proposed by D'Ambrosio, refers to techniques in the broadest sense, including arts. In fact, art plays an important role in the ethnomathematical programme.

4. The ethnomathematical research programme has proliferated worldwide. Thus, we can see studies dealing with mathematics in sugar cane farming (Abreu, 1993; Regnier, 1994). Duarte (2003) addresses the 'world of construction', for instance the mixing of mortar (sand, cement, water). Giongo (2001) analyses the practice of shoemakers. See also Fernandes (2002, 2004). In Brazil researchers and practitioners have struggled with the problems of dealing with hybrid forms of knowledge that characterise the life conditions of many groups of Indians (see, for instance, Amancio, 1999; Scandiuzzi, 2000, 2004). Knijnik (1999) addresses the education for landless people in Brazil. Recently the perspective of street children has been addressed by Mesqita (2004) by investigating the notion of space. Education of indigenous people in Brazil has been addressed by Ribeiro & Ferreira (2004) and Silva (2006) , while the overall ethnomathematical approach has been addressed by Barton (2004).

5. See also Skovsmose (2005).

6. See also Steentoft (2005).

7. It is common that Indians, besides their own name, use a Latin name in situations where they prefer a degree of anonymity. The names used here are their chosen Latin names.

8. In Alrø, Skovsmose and Valero (2005) we have discussed the notion of landscape of learning for the investigation of mathematics education in multicultural settings.

9. The Δ refers to the discriminate playing a role in solved quadratic equations.

10. The universities start at the end of February or at the beginning of March, and during December and January the students, who try to enter universities, go to tests here and there. Naturally, they have to try several universities, as the results of the test are only published later. It could well be that one is not successful at any of the tests (Thus, in many of the subjects the State University of São Paulo only accepts 1 student out of between 20 and 35 students. The competition among the attractive universities is high). If one does not pass any of the entry tests, one has to study an extra year in order to be better prepared. The tests appear horrifying. All kinds of topics are mixed, and the mathematics tasks presuppose not only a deeper understanding of mathematics, but also fingertip knowledge about a variety of details. If astudent should find that a sabbatical year after the gymnasium would be appropriate; there is no chance of coping with the test. One has to enter the test with all the fingertip knowledge necessary. As a consequence, university students in Brazil are much younger than, say, in Denmark. In Brazil one has to enter (or try to enter) immediately after finishing the upper secondary school.

11. For an analysis of the school mathematics tradition and alternative forms of organisation, see Alrø and Skovsmose (2002).
12. Naturally, an answer to such a question reflects also what the students might associate with the words. And it was suggested to the inter-viewer not to use the word mathematics. This might provide some "limited" associations. Therefore the words numbers, calculations, count, estimations were used.
13. It is obvious that the student answers are influenced by one another.
14. For a discussion of intrumentalism in learning mathematics, see Mellin-Olsen (1977). See also Mellin-Olsen (1987).
15. See, for instance, Civil and Andrade (2002).
16. Alan Bishop (1990) gives the example of exercises about cricket scores as well as about the speed of the escalator in Holborn presented for black students in Tanzania, during the colonial times.
17. One could also consider a historical significance. This form of significance has been described by Arthur Powell (2002). He shows how an awareness of the African roots of the Rhind Papirus provides a new significance for the teaching of Afro-American students in New York.

ACKNOWLEDGEMENTS

This chapter makes part of the research project "Learning from diversity." funded by The Danish Research Council for Humanities and Aalborg University.

REFERENCES

Abreu, G. (1993). The relationship between home and school mathematics in a farming community in rural Brazil. Doctoral dissertation. Cambridge: Cambridge University.

Abreu, G., Bishop, A. J., & Presmeg, N. C. (Eds.) (2002). *Transitions between contexts of mathematical practices.* Dordrecht; Boston: Kluwer Academic Publishers.

Alrø, H. & Skovsmose, O. (2002). Dialogue and learning in mathematics education: Intention, reflection, critique. Dordrecht: Kluwer.

Alrø, H., Skovsmose, O., & Valero, P. (2005). *Culture, diversity and conflict in landscapes of mathematics learning.* Paper presented at the CERME 4, Sant Féliu de Gixols. http://cerme4.crm.es/Papers%20definitius/10/wg10listofpapers.htm

Amancio, C. N. (1999). Os Kanhgág da Bacia do Tibagi: Um estudo etnomatemático em comunidades indígenas. Master thesis. Rio Claro: University of São Paulo State.

Barton, B. (2004). Dando sentido à etnomatemática: Etnomatemática fazendo sentido. In J. P. M. Ribeiro, M. do C. S. Domite & R. Ferreira, R. (Eds.). *Etnomatemática: Papel, valor e significado* (39–74). São Paolo: Zouk.

Bishop, A. J. J. (1988). Mathematical enculturation: A cultural perspective on mathematics education. Dordrecht: Kluwer.

Bishop, A. J. (1990). Western mathematics: The secret weapon of cultural imperialism. *Race and Class, 32*(2), 51–65.

Chang, H. (1999). Re-examining the rethoric of the "cultural border." *Electronic magazine of multicultural education, 1*(1), http://www.eastern.edu/publications/emme/1999winter/index.html.

Civil, M. & Andrade R. (2002). Transitions between home and school mathematics: Rays of hope amidst the passing clouds. In G. de Abreu, A. Bishop & N. C. Presmeg (Eds.), *Transitions between contexts of mathematical practices* (149–169). Dordrecht: Kluwer.

D'Ambrosio, U. (1994). Cultural framing of mathematics teaching and learning. In R. Biehler, R. W. Scholz, R. Strässer and B. Winkelmann (Eds.), *Didactics of mathematics as a scientific discipline* (443–455). Dordrecht: Kluwer.

D'Ambrosio, U. (2001). *Etnomatemática: Elo entre tradiçiõs e a modernidade*. Belo Horisonte (Brazil): Autêntica.

Duarte, C. G. (2003). *Etnomatemática, currículo e práticas sociais do 'mundo da constução civil'*. Master thesis. São Leopoldo: Universidade do Vale do Rio dos Sinos.

Fernandes, E. (2002). The school mathematics practice and the mathematics of a practice not socially identified with mathematics. In P. Valero & O. Skovsmose (Eds), *Proceedings of the Third International Mathematics Education and Society Conference* (90–93). Copenhagen, Roskilde, Aalborg: Centre for Research in Learning Mathematics, Danish University of Education, Roskilde University and Aalborg University.

Fernandes, E. (2004). Aprender Matemática para Viver e Trabalhar no Nosso Mundo [Learning Mathematics to live and work in our world] Doctoral dissertation to be published by APM. Lisboa.

Gerdes, P. (1996). Ethnomathematics and mathematics education. In A. Bishop, K. Clements, C. Keitel, J. Kilpatrick, & C. Laborde (Eds.), *International Handbook of Mathematics Education* (909–944). Dordrecht: Kluwer.

Giongo, I., M. (2001). *Educação e produção do calçado em tempos de globalização: Um estudo etnomatemático*. São Leopoldo (Brazil): Universidade do Vale do Rio dos Sinos. (Master thesis.)

Knijnik, G. (1998). Ethnomathematics and political struggles. *Zentralblatt für Didaktik der Mathematik, 30*(6), 188–194.

Knijnik, G. (1999). Ethnomathematics and the Brazilian landless people education. *Zentralblatt für Didaktik der Mathematik, 31(3)*, 188–194.

Knijnik, G. (2002a). Two political facets of mathematics education in the production of social exclusion. In P. Valero & O. Skovsmose (Eds.) (2002), *Proceedings of the Third International Mathematics Education and Society Conference* (357–363). Copenhagen, Roskilde, Aalborg: Centre for Research in Learning Mathematics, Danish University of Education, Roskilde University and Aalborg University.

Knijnik, G. (2002b) Ethnomathematics, culture and politics of knowledge in mathematics education. *For the Learning of Mathematics, 22*(1),11–15.

Knijnik, G. (2002c) Curriculum, culture and ethnomathematics: The practices of 'cubagem of wood' in the Brazilian landless movement. *Journal of Intercultural Studies, 23*(2), 149–166.

Kvale, S. (1996). Inter-views: An introduction to qualitative research inter-viewing. Thousand Oaks, CA: Sage Publications.

MacDonald, R. B., & Bernardo, M. C. (2005). Reconceptualizing diversity in higher education: Borderlands research program. *Journal of Developmental Education, 29*(1), 2–8, 43.

Mellin-Olsen, S. (1987). *The politics of mathematics education.* Dordrecht: Reidel.

Mesquita, M. (2004). O conceito de espaco na cultural de criança em situatição de rua: Um estudo etnomatemático. In J. P. M. Ribeiro, M. do C. S. Domite & R. Ferreira, R. (Eds.), *Etnomatemática: Papel, valor e significado* (125–136). São Paolo: Zouk.

Powell, A. (2002): Ethnomathematics and the challenges of racism in mathematics Education. In P. Valero & O. Skovsmose (Eds.) (2002), *Proceedings of the Third International Mathematics Education and Society Conference* (15–28). Copenhagen, Roskilde and Aalborg: Centre for Research in Learning Mathematics, Danish University of Education, Roskilde University and Aalborg University.

Powell, A. & Frankenstein, M. (Eds.) (1997). *Ethnomathematics: Challenging eurocentrism in mathematics education.* Albany: State University of New York Press.

Regnier, N. M. A. (1994). Ajusta medida: Um estudo das competências matemáticas de trabalhadores da cana-de-açúcar do nordeste do Brasil no domínio da medida. Doctoral dissertation. Paris: Université Rene Descartes—Paris V.

Ribeiro, J. P. M., Domite, M. do C., S. & Ferreira, R. (Eds.) (2004). *Etnomatemática: Papel, valor e significado.* São Paolo: Zouk.

Ribeiro, J. P. M. & Ferreira, R. (2004). Educacao escolar indídiga e etnomatemática: Um diálogo necessário. In J. P. M. Ribeiro, M. do C. S. Domite & R. Ferreira, R. (Eds.). *Etnomatemática: Papel, valor e significado* (149–160). São Paolo: Zouk.

Scandiuzzi, P., P. (2000). Educação Indígena x Educação Escolar Indígena: Uma relação etnocida em uma pesquisa etnomatemática. Marilia (Brazil): State University of São Paulo. Doctoral thesis.

Scandiuzzi, P. P. (2004). Educação Metemática Indídiga: A Constitução do Ser entre os Saberes e Fazeres. In M. A. V Bicudo & M. C. Borba (Eds.), *Educação Matemática: Pesquisa em Movimento* (186–197). São Paulo: Cortez Editoria.

Silva, A. A.. (2006). *A organização espacial A`UWe—Xavante: Um olhar qualitativo sobre o espaço.* Rio Claro (Brazil): Universidade Estadual Paulista, Institutio de Geociência e Ciências Exatas.

Skovsmose, O. (2005). Foregrounds and politics of learning obstacles. *For the Learning of Mathematics, 25*(1), 4–10.

Steentoft, D. (2005). Research as an act of learning: Exploring student backgrounds through dialogue with research participants. *Proceedings from CERME-4,* Barcelona 2005.

CHAPTER 14

ICELAND AND RURAL/URBAN GIRLS-PISA 2003 EXAMINED FROM AN EMANCIPATORY VIEWPOINT

Olof Bjorg Steinthorsdóttir
University of North Carolina-Chapel Hill, USA

Bharath Sriraman
The University of Montana, USA

ABSTRACT

Scholarly research related to gender and mathematics is not as frequently published as it was in the 1980s and the 1990s. In Lubienski's (2000) survey of Mathematical Education Research from 1982 to 1998 there are 367 publications in Journal for Research in Mathematics Education (JRME) and 385 publications in Educational Studies in Mathematics (ESM) about gender. This gives us approximately 21 publications a year in JRME and 22 publications a year in ESM. We did a search of publications about gender in JRME and ESM from 1999 to 2005 (or today) and saw a very different picture. Over this period 14 publications were in JRME and 17 publications in ESM, which gives approximately 4 publications a year in JRME and 3 publications a year in ESM.

International Perspectives on Social Justice in Mathematics Education, pages 231–244

So what do these numbers tell us about the status of research about gender and mathematics? Does this mean that the gender gap has been closed? If so, for whom is that true? Does it mean that we don't have to worry about gender differences in mathematics any more? And if it is true, is it certain that it will sustain itself without any follow up? Finally, why are there still differences in women entering fields such as mathematics and physics?

A BRIEF SURVEY OF GENDER DIFFERENCES

According to Albert Bandura's (1977) persistence theory self-efficacy is positively related to persistence. In other words persistence on a (math) problem in spite of frustrations is more likely to lead to a solution/success (Brown, Lent & Larkin, 1989; Schunk, 1985). Low self-efficacy in females has been attributed to low parental expectancies and sexual stereotyping in the attitudes of teachers and male students in school. The literature in gender studies suggests that society as whole believes that females are less mathematically capable than men (Aiken, 1974; Burton, 1979; Fenemma & Sherman, 1977, 1978). The findings were also not different for gifted girls (Benbow & Stanley, 1980; Cramer, 1989; Eccles, 1985). Females are particularly vulnerable to the stereotype that "girls just can't do math" and when women go onto courses like calculus they fare less well than men who have shown equal promise up to that point (Fennema & Sherman, 1978). *Arguably the aforementioned literature is a bit outdated, which leads one to question whether the situation is different today?* At ICME 10 (International Congress of Mathematics Education) held in Copenhagen in 2004, Topics Study Group 26 was about gender and mathematics education and 15 papers were presented. Two studies from Scandinavia showed interesting results about gender differences still in existence. In particular, a study from Sweden with 9th and 11th grade students showed that students still viewed mathematics as a male domain (Brandell, Nystrom, & Sundqvist, 2004). Another study from Finland reported that teachers held different beliefs about girls and boys in their classroom, believing that girls tended towards routine procedures whereas boys use their power of reasoning (Soro, 2004). These findings suggest that not much has changed in terms of society's dominant conceptions of mathematics.

A study from the US indicated that girls' self-confidence was an important factor when it came their participation in mathematics. In this study 121 middle school girls took part in a 5 day residential summer camp about mathematics. The population distribution was 69% white, 10% American Indian, 7% Hispanic, 4% Asian, 2% black and 8% bi-racial. The summer camp had a positive impact on their achievement. These researchers suggested that the girls' increase in their self-confidence accounted for their achievement (Wiest, 2004). Again the implication we draw from this study is that girls' self-esteem was still perceived as an important factor when it came to their career choices and higher education.

Another study from South Africa reported gender differences in attitudes toward mathematics are in favor of males. Girls reasons for taking math course was external while the boys reasons were internal. Girls reported that math was difficult whereas boys did not think so. And finally girls did not think that math was particularly useful in their home environment whereas boys said it was useful because it was in their environment. These authors suggested that these fundamentally different views were due to differentiated socialization processes (Mahlomaholo & Sematle, 2004).

An Australian study reported gender differences in the use of technology (computers and hand-held technology) in mathematics classes. Males were more likely to believe that technology had a positive effect on mathematical learning. In addition, teachers perceived that males had more suitable characteristics to benefit from using technology to advance in mathematics. Compared to females, males were also more prepared to take risks and have a go at using the software. Is mathematics is doomed to be considered a male domain? (Forgasz, 2004).

A study from Iran which analyzed the University Entrance Exam reported that females were less interested in mathematics than their male counterparts. Also male students are more successful in Mathematics and Physics, and the acceptance rate based on the Exam was in favor of males (in 2003 approximately 64% were male) (Pourkazemi, 2004).

Becker and Rivera (2004) presented a synthesis of perspectives used to investigate gender and mathematics in different countries (from the gender working group for the last several meetings of PME-NA and PME international). There synthesis suggests that four perspectives are present in the research on gender and mathematics. They label them (1) Predict, (2) Understand, (3) Emancipate, and (4) Deconstruct. The first perspective Predict falls under the umbrella of a positivist view and many studies conducted in the '70s and '80s fall in this category. The second perspective, Understand, are studies that attempt to make sense of the reality of gender and math without changing the social environment. We would argue that many of the studies looking at gender and mathematics in the 90's fall under this category. The third perspective Emancipate is research where gender is not seen as an isolated variable but is intertwined with race, class, ethnicity, and culture. Finally in the fourth perspective, Deconstruct, gender is viewed as performances and is subject to social construction and the goal is to deconstruct common beliefs. Their findings suggest that even today not many studies about gender and mathematics fall under the third and fourth perspectives.

PISA (PROGRAMME FOR INTERNATIONAL STUDENT ASSESSMENT) 2000 AND 2003

Despite the common belief (in many western countries) that the gender differences in mathematical achievement has been bridged, PISA, in addition to various presentations at ICME 10 provided documented statistically significant gender differences in achievement in favor of boys both in the year 2000 and 2003. In the year 2000 statistically significant gender difference in achievement were found in 29 countires of the 41 participating countires. In year 2003 statistically significant gender differences in achievement were found in 27 countires of the 41 participating countires. The only country in PISA 2003 which had statistically significant gender differences in achievement in favor of girls was Iceland. In the following sections we will give a brief review of the PISA study followed by a closer look at the Icelandic data.

PISA—INTRODUCTION

In today's society the prosperity of a country is largely dependent on their human capital and how well individuals can advance their knowledge and skill in a rapidly changing world. In 1997 The Programme of Economic Cooperation and Development (OECD) launched the Programme for International Student Assessment (PISA) to develop an international study. The cross-national comparison on students performance could provide contries with information to judge their strengths and weaknesses and to monitor progress. PISA seeks to measure how well students at the age of 15 are prepared to meet the challenges of today's knowledge societies.

The key features of PISA have been:

- its policy orientation, with design and reporting methods determined by the need of governments to draw policy lessons
- the innovative "literacy" concept that is concerned with the capacity of students to apply knowledge and skills in key subject areas and to analyze, reason and communicate effectively as they pose, solve and interpret problems in a variety of situations
- its relevance to lifelong learning, which does not limit PISA to assessing students' curricular and cross-curricular competencies but also asks them to report on their own motivation to learn, their beliefs about themselves and their learning strategies;
- its regularity, which will enable countries to monitor their progress in meeting key learning objectives

- its breadth of geographical coverage and collaborative nature, with the 49 countries that have participated in a PISA assessment so far and the 11 additional countries that will join the PISA 2006 assessment representing a total of one third of the world population and almost nine-tenths of the world's gross domestic product (GDP) (OECD, 2003, p. 20).

PISA measures students´ preformance in literature, mathematics, and science and is conduceted in stages. The first stage was conducted in 2000 with literature was the main focus. The second stage of the sudy was conducted in 2003, where mathematics was the main focus. The third stage is 2006, where the primary focus will be science. In 2009 the circle will start again, with the main focus on reading.

ITEM DESIGN, ANALYSIS AND SCALES

PISA measures student's mathematical knowledge and skills, which are assessed according to the following three dimensions.

1. The mathematical content to which different problems and questions relate
2. The processes that need to be activated in order to connect observed phenomena with mathematics and then to solve the respective problems
3. The situations and contexts that are used as sources of stimulus materials and in which problems are posed (OECD, 2003, p. 38)

The mathematical content area assessed in PISA is built on a consensus among OECD countries and appropriate for international comparison. The content areas are:

- *Space and shape* relates to spatial and geometric phenomena and relationships, often drawing on the curricular discipline of geometry. It requires looking for similarities and differences when analyzing the components of shapes and recognizing shapes in different representations and different dimensions, as well as understanding the properties of objects and their relative positions.
- *Change and relationships* involves mathematical manifestations of change as well as functional relationships and dependency among variables. This content area relates most closely to algebra. Mathematical relationships are often expressed as equations or inequalities, but relationships of a more general nature (*e.g.,* equivalence,

divisibility and inclusion, to mention but a few) are relevant as well. Relationships are given a variety of different representations, including symbolic, algebraic, graphic, tabular and geometric representations. Since different representations may serve different purposes and have different properties, translation between representations is often of key importance in dealing with situations and tasks.

- *Quantity* involves numeric phenomena as well as quantitative relationships and patterns. It relates to the understanding of relative size, the recognition of numerical patterns, and the use of numbers to represent quantities and quantifiable attributes of real-world objects (counts and measures). Furthermore, quantity deals with the processing and understanding of numbers that are represented in various ways. An important aspect of dealing with quantity is quantitative reasoning, which involves number sense, representing numbers, understanding the meaning of operations, mental arithmetic and estimating. The most common curricular branch of mathematics with which quantitative reasoning is associated is arithmetic
- *Uncertainty* involves probabilistic and statistical phenomena and relationships that become increasingly relevant in the information society. These phenomena are the subject of mathematical study in statistics and probability (OECD, 2003, p. 38–39)

The items had a variety of formats where various competencies were required. They include *thinking and reasoning, argumentation, communication, modeling, problem posing and solving, representation, and using symbolic, formal and technical language and operation.* The problems were then organized in three competency clusters. The first being *reproduction cluster,* the second being *connection clusters,* and finally *reflection clusters.* The tasks were set in different context varying in the degree of distance between the student and the situation, (1) personal situation, (2) educational or occupational situation, (3) public situation relating to the community, and (4) scientific situation.

The problems varied in formats. The items were categorized into (1) multiple choice, (2) complex multiple choice, (3) closed constructed response, (4) open constructed response, and (5) short response. According to the characteristics of each task and which competency they address the tasks were labeled from one to six according to difficulties, one being the easiest. Students were categorized according to these six proficiency levels depending on their scores and which problem they could solve (see Figure 14.1).

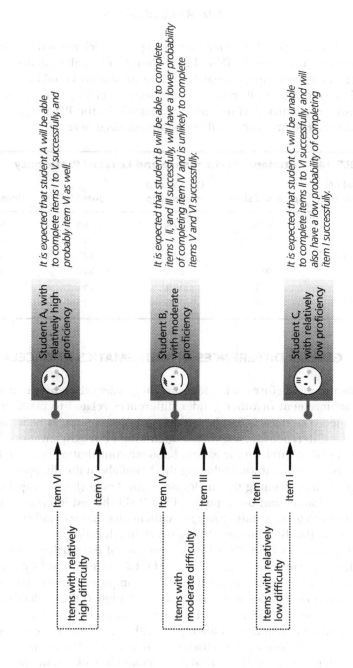

Figure 14.1 The relationship between items and students on a proficiency scale (OECD, 2003, p. 45).

OVERALL RESULTS

Overall students' achievement according to proficiency levels in each concept area is presented in Table 14.1. Around 90% of all students could solve level I problems. Approximately 50% of all students could solve level III, and around 5% of all students could solve level VI problems. Analysis of individual country performances is available in the PISA report (http://www.pisa.oecd.org) but it will not be summarized here.

TABLE 14.1 Students Performance and Level of Proficiency

Level of proficiency	Space and shape	Change and relationship	Quantity	Uncertainty
I	87%	87%	88%	90%
II	71%	73%	74%	75%
III	51%	54%	53%	54%
IV	30%	32%	31%	31%
V	15%	15%	13%	13%
VI	5%	5%	4%	4%

GENDER DIFFERENCES IN MATHEMATICS AND ICELAND

As mentioned before studies focusing on gender differences in mathematical achievement or other gender differences related to mathematics has declined considerably for the last 10 year. Also, a popular belief is that the gender difference favoring male students does not exist any more. Whether the lack of research on the issue at hand is a contributing factor to this common belief is unknown. Following this belief about the disappearing differences, voices claiming that males were now being shortchanged in school became louder and more public. PISA 2003 showed interesting results in relations to gender differences in mathematics that contradict the popular discourse that boys are on a losing streak in education.

According to PISA 2003, in just over half of the participating countries males outperformed females, or in 17 OECD countries and 4 partner countries.[1] In addition, in mathematics and computer science, gender differences favoring males remains persistently high (OECD, 2003). Looking closer at the graduating rate of females in different subject areas, the average number of females graduating in mathematics and computer science is only about 30% of total graduation. Interestingly, despite the reverse gender differences in mathematical achievement in Iceland, the proportion of females graduating with a postsecondary degree or higher in mathematics and computer sciences is just around 20%. For comparison postsecondary graduation of women in humanities, arts and sciences is around 70& within OECD countries but reaches 80% in Iceland (OECD, 2004).

In Iceland, as mentioned before, there were significant gender differences in mathematics achievement in favor of girls. Dividing Iceland into two regions, Reykjavik metropolitan area and rural area, significant gender differences in achievement was only found in rural Iceland. For further analysis Iceland was divided into 9 regions, 2 of them being Reykjavik metropolitan area and 7 rural areas. The largest difference was found on the south coast of Iceland with the point difference being 30 points. The lowest difference outside the Reykjavik area was 12 points found at the east fjords. Other five areas point differences was 17, 20, 25, and 26 points. As mentioned above the gender differences in the Reykjavik metropolitan area was not significant, or on an average 4 points (Olafsson, Halldorsson, & Bjornsson, 2006).

PISA 2003 categorized mathematical performance into 6 levels, level 0 being low proficiency and level 6 being high proficiency. In general the results show that more males than females reached level 6 or, 7% of males and 4% of females. In Iceland around 4% of all students performed at level 6. The gender differences in number of students performing at level 6 differ by regions. In Reykjavik and surrounding areas a greater portion of males reached level 6 compared to females. On the other hand in rural areas more females reached level 6 than males (Olafsson, Halldorsson, & Bjornsson, 2006). It is interesting that the difference at level 6 was less than the differences at other levels. At the other end of the spectrum, that is students at Level 0 and Level 1, more boys than girls were categorized at these levels in all regions, 18% and 11% respectively.

When the four areas of mathematics that were tested are analyzed according to gender it can be seen that the gender differences are not the same across areas. The largest difference was found in Quantities, followed by Space and Shape and Change and Relationship. The smallest difference was found in the category Uncertainty. It is also interesting that Iceland scored highest in the uncertainty category of all the countries (Olafsson, Halldorsson, & Bjornsson, 2006). Despite this unusual gender differences in favor of girls found in Iceland, Icelandic girls are not different from other girls in the study when it comes to math anxiety, and mathematical confidence, there the gender differences are in favor of boys (Olafsson, Halldorsson, & Bjornsson, 2006).

DISCUSSION AND REFLECTIONS

Some thoughts and ideas have been tossed around in the attempt to explain these unusual results. We will now share some of these ideas and add some of our own ideas.

One of the more popular explanations is so called "jokkmokk" effect. To explain it simply, jokkmokk effects refer to this "phenomena" of females outperforming males academically in rural areas. It suggest that the environment, such as the labor market, prevent males to see value in academic education, on the contrary the same environment encourage females to do well in school in the hope of achieving some status in their future or leave their hometown in search for a "better" life. Applying this idea to the Icelandic situation it probably has some effect but to believe it is the answer is naïve. It is true in some rural areas in Iceland males can be financially successful without a post secondary degree. On the other hand most traditional female jobs today require college degrees, such as nurse, teacher, or bank teller.[2]

Another explanation could be related to school environment and the gendered discourse that takes place among teenagers. A study in Iceland reported on interesting gender differences in what is accepted discourse among teenagers in Iceland (Magnusdottir, 2005). Their findings imply that it is accepted that girls work hard to get good grades and in fact it is expected of them to do so if they want to get good grades. For boys on the other hand it is not the case. The common belief is that boys do not have to study, they get good grades anyway. One can argue that most individuals, females or males, have to study to achieve good grades. With that in mind and the PISA results the teenage boys are then more likely to achieve lower scores than teenage girls. That is, if it is not "cool" for the teenage boys to study than one can expect that only few teenage boys will achieve high scores (assuming that most teenage boys are influenced by the dominant discourse in their peer group). But the question remains how does this argument explain the lack of gender difference in Reykjavik metropolitan area? Also the fact remains that a higher proportion of boys in the Reykjavik metropolitan area reached level 6. Are self efficacy and stereotyping vulnerability in mathematics contributing factors?

Related to the gendered discourse explanation is to examine if the classroom is a feminine environment and therefore less suited for boys. This question has been found within the circle of researchers and laymen that have been questioning the status of boys in primary and secondary education. Two Icelandic women Berglind R. Magnusdottir and Thordgerdur Einarsdottir (2006) make a compelling argument that rejects this notion in Iceland, one being the structure of the academic system from a historical point of view. Even though schools today have more female teachers and included more of what would be categorized as "feminine" trades, such as caring, cooperation and shared management, the "masculine" trades still have strong hold in the foundation of the educational system, such as teacher-center pedagogy, lectures, and individual work. One of the more "amusing" arguments that they provide is an Icelandic study that show that boys complain more that they do not feel so good in school and this complaint gets louder with age (Jonsdottir et. al, 2002). What is interesting about this is that in Iceland, as in many countries, the number of male teachers increases in secondary school. Maybe one can argue that with more male teachers, male students' feelings worsen. Finally, when the use of the special education budget is examined, proportionally more is spent on male students and in addition male students gain more from the special education that is offered in schools. With that being said and in all seriousness related to the number of male students in special education, we do think it is an issue that is worthy of more research.

The last idea that we will present, and the most interesting one to us, is the correlation between students' reading comprehension and mathematical achievement. An analysis of the achievement of Norwegian and Swedish students show high correlation[3] between the two, that is, high score in reading comprehension correlates with high score in mathematics (Roe & Taube, 2006). In addition the Norwegian and Swedish analysis shows that the strength of the correlation is greater on items that called for "more" complex mathematical understanding, such as reproduction and open constructed answers.

The gender differences in reading comprehension in favor of girls were the largest in Iceland of all the participating countries and were significant in all regions of Iceland. When the data was analyzed by controlling for reading comprehension the males scored little higher than females. That is, given the same level of reading ability one could predict that female would achieve lower scores than males (Olafsson, Halldorsson, & Bjornsson, 2006). These results are interesting and even though the gender differences in the urban area was slightly lower than the difference in the rural area it does not fully explain the lack of gender differences in mathematics in the Reykjavik area. The correlation between reading and mathematical achievement can potentially provide partial answers in the search for explanation and needs to be studied further.

Maybe the last question to ask is if these results are reliable or a flukish-one time results. Olafsson, Halldorsson, & Bjornsson (2006) look at the Icelandic National Mathematics Test scores from 1994 to 2004. According to their analysis the gender differences in mathematics in favor of girls has been measured all the years mentioned. On the other hand the differences between scores of urban and rural students are not consistent, that is over these 10 years the gender differences vary across regions each year. It is important to mention that The Icelandic National Mathematics Test scores for year 2003 mirrors the outcome of PISA 2003. We can then argue that the PISA result has some merits and deserve further research.

With all this said the question about the Icelandic "phenomena" remains mostly unanswered at this time. One thing that is clear to us though is that poor performance of boys in mathematics is *not* because they are receiving lesser quality mathematical instruction in school like the picture some media and politicians are trying to paint. As so often is the case, in the search for answers, before the answer is "found" more questions are generated with each step. We will therefore end this chapter by posing few questions that we feel compelled to ask or we interpret what the data suggests.

- If boys' reading comprehension score were higher or at the same level as girls would they do better than girls? In mathematics? In other subjects?
- Should the schools focus on boys reading comprehension more?
- Is the "problem" because of the mathematics curriculum or the teaching of mathematics?
- Are boys falling behind in mathematics because "doing" school mathematics calls for more reading and writing than it used to?

Our next step is indeed to search for some concrete answers!

NOTES

1. 30 OECD countries participated and 11 partner countries.
2. In today society jobs such as bank teller is often occupied by people with some business degrees.
3. Correlation coefficient of 0.57

REFERENCES*

Aiken, L.R. (1974). Affective variables and sex differences in mathematical abilities. Paper presented at the annual meeting of the American Educational Research Association, Chicago.

Bandura, A. (1977). *Social Learning Theory.* New York: General Learning Press.

Bandura, A. (1977). Self-efficacy: Toward a unifying theory of behavior change. Psychological Review, 84, 191–215.

Becker. J. R. & Rivera, F. (2004). *Emerging perspectives of research on gender and mathematics: A global synthesis.*Paper presented at the quadrennial conference of the International Congress on Mathematical Education (ICME 10), Copenhagen, Denmark.

Benbow, C.P & Stanley, J.C. (1980). Sex differences in mathematics ability, *Science, 210,* 1262–1264.

---

* All Icelandic titles are translated by Olof Bjorg Steinthorsdottir.

Brandell, G., Nyström,P., & Sundqvist, C. (2004). *Mathematics—a male domain*. Paper presented at the quadrennial conference of the International Congress on Mathematical Education (ICME 10), Copenhagen, Denmark.

Brown, S. D., Lent, R. W., & Larkin, K. C. (1989). Self-efficacy as moderator of scholastic aptitude-academic performance relationships. *Journal of Vocational Behavior*, 35, 64–75

Burton, G.M. (1979). Regardless of sex, *Mathematics Teacher*, 261–270.

Cramer, R.H. (1989). Attitudes of gifted boys and girls towards math: a qualitative study. *Roeper Review, Vol. 11*, No.3, 128–131.

Eccles, J.S. (1985). Why doesn't Jane run? Sex differences in educational and occupational patterns. In F.D. Horowitz & M.O'Brien (Eds.), *The gifted and talented: developmental perspectives (pp. 251–295)*, Washington D.C, American Psychological Association.

Fennema, E. & Sherman, J.C. (1977) Sex related differences in mathematical achievement, spatial visualization and affective factor, *American Educational Research Journal*, 51–71.

Fennema, E., & Sherman, J. A. (1978). Sex-related differences in mathematics achievement, spatial visualization and affective factors: A further study. *Journal for Research in Mathematics Education*, 9, 189–203.

Forgasz, H. (2004). *Computers for mathematics learning and gender stereotypes*. Paper presented at the quadrennial conference of the International Congress on Mathematical Education (ICME 10), Copenhagen, Denmark.

Jonsdottir, S. N., Bjornsdottir, H. H., Asgeirsdottir, B. B, & Sigfusdottir, I. D. (2002). *Börnin í borginni. Liðan og samskipti í skóla, félagstarf og tómstundir og vímuefnaneysla. Könnun meðal nemenda í 5.–10. bekk grunnskóla Reykjavíkur vorið 2002* [City children. Feelings and communication in school, after school activity and drug use. Research among students in 5–10 grade in compulsory schools in Reykjavik in spring 2002]. Reykjavik, Iceland: Rannskoknir og greining.

Olafsson, R. F., Halldorsson, A. M., & Bjornsson, J. K. (2006). Gender and the urban-rural differences in mathematics and reading: An overview of PISA 2003 results in Iceland. In J. Mejding & A. Roe (Eds.), *Northern lights on PISA 2003—a reflection form Nordic countries*, (185–198). Copenhagen, Denmark: Nordic Council of Ministers.

Lubienski, S. T. & Bowen, A. (2000). Who's counting? A survey of mathematics education research 1982–1998. *Journal for research in mathematics education*, 31, 626–633.

Magnusdottir, B. R. (2005) "Ég veit alveg fullt of hlutum en ...". Hin kynjaða greindarorðræða og birtingarmyndir hennar meðal unglinga í bekkjadeild ["I know lots of things but ...". The gendered discourse among classroom teenagers]. In A. H. Jonsdottir, S. H. Larusdottir, & Th. Thordardottir (Eds.), *Kynjamyndir i skolastarfi*. (pp. 151–172) Reykjavik, Iceland: Iceland University of Education.

Magnusdottir, B. R. & Einarsdottir, Th. (2005). ER grunnskólinn kvenlæg stofunun? [Is the primary and secondary school a feminine institute?]. In A. H. Jonsdottir, S. H. Larusdottir, & Th. Thordardottir (Eds.), *Kynjamyndir i skolastarfi*, (pp. 173–198) Reykjavik, Iceland: Iceland University of Education.

Mahlomaholo, S. & Sematly, M. (2004). *Gender differences and black learners' attitudes towards mathematics in selected high schools in South Afrika.* Paper presented at the quadrennial conference of the International Congress on Mathematical Education (ICME 10), Copenhagen, Denmark.

Pourkazemi, M. H. (2004). *In the name of God. Gender and mathematics.* Paper presented at the quadrennial conference of the International Congress on Mathematical Education (ICME 10), Copenhagen, Denmark.

Programme of Economic Co-operation and Development (OECD). (2003). *Learning for tomorrow's world: First results from PISA 2003.* Paris, France: The Programme of Economic Co-operation and Development.

Programme of Economic Co-operation and Development (OECD). (2004). *Education at glance: OECD indicators 2004.* Paris, France: The Programme of Economic Co-operation and Development.

Roe, A. & Taube, K. (2006). How can reading ability explain differences in maths performances? In J. Mejding & A. Roe (Eds.), *Northern lights on PISA 2003—a reflection form Nordic countries,* (129–142). Copenhagen, Denmark: Nordic Council of Ministers.

Schunk, D. H. (1985). Self-efficacy and classroom learning. *Psychology in Schools,* 22, 208–223.

Soro, R. (2004). *Teachers' beliefs about girls and boys equity in mathematics.* Paper presented at the quadrennial conference of the International Congress on Mathematical Education (ICME 10), Copenhagen, Denmark.

Wiest. L. R. (2004). *The critical role of an informal mathematics program for girls.* Paper presented at the quadrennial conference of the International Congress on Mathematical Education (ICME 10), Copenhagen, Denmark.

ABOUT THE CONTRIBUTORS

Helle Alrø is Professor at Aalborg University, Department of Communication. She has a special interest in dialogic learning processes in educational as well as organizational contexts. Her resent research is concerned with conflict and conflict management as a means for learning. She has published books and articles on interpersonal communication in helping relationships, i.e. on communication in the mathematics classroom. Most of the latter work has been carried out in collaboration with Ole Skovsmose.

Miriam Amit is a Professor of Mathematics Education and Head of the Department of Science and Technology Education in the Ben-Gurion University of the Negev. She completed her Bachelors and Masters degrees in the Technion—the Israeli Institute of Technology, and her PhD in the Department of Mathematics and Computer Sciences in Ben-Gurion University. For 12 years, she was the Chief Superintendent of Mathematics Education for the State of Israel, in charge of K–12 curriculum design and implementation, teacher development and national assessment. In the latter field she also worked as a consultant for major international research institutions. Among her interests are: social and cultural aspects of mathematics education, which include ethno-mathematics and gender; alternative assessment methods and its connection to problem solving, learning and instruction; and last but not least, the cultivation and promotion of excellence among mathematically promising students. To further this final goal, Professor Amit founded the "Kidumatica Youth Club", which since its founding 8 years ago has brought excellence to thousands of children.

Kristín Bjarnadóttir studied physics and mathematics at the University of Iceland and completed her M.Sc. degree in mathematics at the University of Oregon, U.S., in 1983. She defended a Ph.D. thesis, Mathematical Education in Iceland in Historical Context—Socio-Economic Demands and Influences, at Roskilde University, Denmark, in February 2006. Her supervisor was Prof. Mogens Niss. Kristín Bjarnadóttir has taught mathematics and

International Perspectives on Social Justice in Mathematics Education, pages 245–251
Copyright © 2008 by Information Age Publishing
245

physics at lower and upper secondary schools in Iceland. Currently she is Associate professor at Iceland University of Education.

Iben Maj Christiansen works in the School of Education and Development, Faculty of Education, University of KwaZulu-Natal, Pietermaritzburg in South Africa. Before that, she was an Associate Professor in Mathematics and Science Education at Aalborg University in Denmark. There, she also served as director of the Centre for Educational Development of University Science, which offered staff development and coordinated action research projects at seven higher education institutions. Iben is involved in research on tacit knowledge in mathematics education and teacher training, in particular in relation to the teaching and learning of calculus for teachers, and developmental work in higher education in South Africa. She is the mother of twin boys, aged 5.

Ubiratan D'Ambrosio is the founder of the ethnomathematical movement. He is the 2005 recipient of the Felix Klein Medal of Mathematics Education, granted by the International Commission of Mathematics Instruction (ICMI), the highest honor in the mathematics education community. His educational background includes a doctorate in Mathematics, Universidade de São Paulo (1963). He is now an Emeritus Professor of the State University of Campinas/UNICAMP, São Paulo, Brazil and Fellow of the AAAS: American Association for the Advancement of Science with the citation "For imaginative and effective leadership in Latin American Mathematics Education and in efforts towards international cooperation." (1983). He also received the "Kenneth O. May Medal in the History of Mathematics", granted by the International Commission on History of Mathematics (2001). His current activities include: Professor at the PUCSP/Pontifícia Universidade Católica de São Paulo, and guest professor at the USP/Universidade de São Paulo and the UNESP/Universidade Estadual Paulista.; and President, Brazilian Society of History of Mathematics/SBHMat.

Michael N. Fried is a lecturer in the Program for Science and Technology Education at Ben Gurion University of the Negev. His undergraduate degree in the liberal arts is from St. John's College in Annapolis MD (the "great books" school). He received his M.Sc. in applied mathematics from SUNY at Stony Brook and his Ph.D. in the history of mathematics from the Cohn Institute of History and Philosophy of Science at Tel Aviv University. His research interests are eclectic and include mathematics teacher education, sociocultural issues, semiotics, history of mathematics, and history and philosophy of education. He is author with Sabetai Unguru of Apollonius of Perga's Conica: Text, Context, Subtext.

Merrilyn Goos is an Associate Professor in the School of Education at The University of Queensland, Australia, where she co-ordinates pre-service and postgraduate courses in mathematics education. Her research has been guided by sociocultural theories of learning in investigating classroom interactions and mathematical thinking, analysing pedagogical issues in introducing educational technologies into mathematics teaching and learning, and studying how communities of practice are established and maintained in secondary mathematics classrooms and teacher education contexts. In 2002–2003 she led a national numeracy research project commissioned by the Australian Government Department of Education, Science and Training that investigated home, school and community partnerships in children's numeracy education. This project integrated multiple disciplinary and methodological perspectives in collecting and analysing survey, interview and case study data from all Australian states and territories.

Brian Greer came to mathematics education with a background in mathematics and psychology, leading to an interest in the relationship between cognitive psychology and mathematics education. After some 30 years in the School of Psychology in Belfast, Ireland, he took a position in mathematics education at San Diego State University, which he left in 2003 to work as an independent scholar in Portland, Oregon. Topics that he has focused on include multiplicative structures, probabilistic thinking, and word problems. More recently, particularly under the influence of Swapna Mukhopadhyay, with whom he collaborates intensively, he characterizes mathematics and mathematics education as human activities that are historically, culturally, socially, and politically situated.

Eric Gutstein is Associate Professor of Curriculum and Instruction at University of Illinois-Chicago. His interests include teaching mathematics for social justice, Freirean approaches to teaching and learning, and urban education. He has taught middle and high school mathematics. Rico is a founding member of Teachers for Social Justice (Chicago) and is active in social movements. He is the author of Reading and Writing the World with Mathematics: Toward a Pedagogy for Social Justice (Routledge, 2006) and an editor of Rethinking Mathematics: Teaching Social Justice by the Numbers (Rethinking Schools, 2005).

Ravin Gustafson, a freelance editor for many years, has been a middle school teacher for the past 4 years. She began as a language arts teacher and has recently transitioned to a dual role: teaching a self-contained classroom of 7th graders and teaching 7th and 8th grade mathematics. Ravin has a lifetime connection to Native America.

Lesley Jolly is an anthropologist lecturing in Behavioural Studies at The University of Queensland, Australia. She has carried out research in cultures as diverse as those of indigenous Australian societies in both remote and urban Australia, engineering and high technology workplaces, suburban community groups and elementary classrooms. The issues she has addressed have included cultural constructions of public and private interaction, gender in engineering, technology in the classroom and in non-professional communities. In 2002–2003 she took part in the national numeracy research project commissioned by the Australian Government Department of Education, Science and Training that investigated home, school and community partnerships in children's numeracy education which is drawn on in this paper.

Libby Knott is an Associate Professor in the Department of Mathematical Sciences at the University of Montana. She has a PhD in mathematics education from Oregon State University and degrees from Wesleyan University and Colorado State University. She has taught at Lewis and Clark College (Portland, OR), SUNY College at Cortland, The University of California, Santa Barbara, Oregon State University and has been on the mathematics faculty at The University of Montana since 1996. Interests include all aspects of undergraduate mathematics teaching and learning. She is an author of the 9–12 Navigating through Geometry book in the NCTM Navigations series. Current research interests include the role of discourse mathematics teaching and learning. She is also involved in the preparation and professional development of middle and high school math teachers in the state of Montana.

Tom Lowrie is Professor and Head of the School of Education at Charles Sturt University, Australia. His research interests are influenced by the way in which students' use spatial reasoning and visual imagery to solve mathematics problems in contexts that are authentically based and in learning situations that are very different to that of traditional classrooms. He is a co-author of Mathematics for children: Challenging children to think mathematically and has published extensively in mathematics education and related disciplines from both psychological and sociology perspectives.

Swapna Mukhopadhyay came to mathematics education with a background in physics and anthropology, together with experience in teaching in an innovative alternative school in her native Calcutta. She is now an Associate Professor at the Graduate School of Education at Portland State University. She regards mathematics as cultural construction, particularly through the lens of ethnomathematics, the mathematical knowledge construction and use of social groups, including children learning mathematics at school and people acting in their everyday lives. She promotes a vision of mathematics education as providing people with tools for critiquing and acting upon issues important in their lives.

Mohammed Abu-Naja received his Ph.D. in mathematics education from Ben-Gurion University of the Negev. He is a lecturer and pedagogical instructor for prospective high school mathematics teachers at Kaye Teachers' Academic College, a lecturer at Achva College, and instructor in a program for student excellence under the auspices of the Technion and Hebrew University. He also worked as a high school teacher and department chairman for 14 years. His M.Sc. concerned the influence of mathematical recreations on attitude and achievement among 10th graders in the Bedouin sector, while his doctoral work concerned how Bedouin students use graphing calculators to learn central concepts in the mathematics curriculum. His research interests include: the use of graphing calculators in mathematics teaching and learning, the development of mathematical thinking, teaching geometry to pre-service teachers.

Tod Shockey is an Assistant professor of Mathematics Education at the University of Maine. Prior to work in higher education he was a secondary level mathematics teacher. His research interest is focused in ethnomathematics.

Ole Skovsmose has a special interest in critical mathematics education. Recently he has published investigations of mathematics in action, students' foreground, globalisation and ghettoising. He is Professor at Aalborg University, Department of Education, Learning and Philosophy. He is member of the editorial boards of Mathematics Education Library (Springer) and Critical Essays in Education (Sense Publisher). He has participated in conferences and given lectures about mathematics education in many different countries. His most recent book is entitled *Travelling Through Education: Uncertainty, Mathematics, Responsibility*, Sense Publishers, The Netherlands. This book can be downloaded in PDF for free at www.sensepublishers.com.

Bharath Sriraman is an Associate Professor of Mathematics at the University of Montana, with eclectic research interests. Bharath, a native of India, lives in Montana byway of the merchant marine; Alaska (B.S in mathematics, University of Alaska- Fairbanks) and Illinois (M.S & PhD in mathematics and mathematics education, Northern Illinois University). He is interested in Cognitive Science; Innovation and Talent Development; History and Philosophy of Mathematics; History of Science; and Mathematics Education. He has published over 100 refereed papers, commentaries, book chapters and book reviews in his areas of interest. Bharath maintains an active research interest in Indo-Iranian Studies and the evolution of human societies. He is more or less fluent in 7–9 languages [English, German, Farsi, Hindi, Kannada, Urdu, Tamil, French and others]. Bharath is the Editor of *The Montana Mathematics Enthusiast*; Associate Editor of Zentralblatt für Didaktik der Mathematik and serves on the editorial boards of seven other journals. He is the Book Reviews Editor of Mathematical Thinking & Learning: An International Journal and Zentralblatt für Didaktik der Mathematik. His interest in social justice originally stems from familiarity with the writings of Karl Marx, Raja Ram Mohan Roy and Vivekananda. Bharath holds active research ties with researchers working in his domains of interest in Australia, Canada, Cyprus, Denmark, Germany, Greece, Iceland, India and Turkey.

Olof Bjorg Steinthorsdottir is currently an Assistant Professor of mathematics education in the School of Education at University of North Carolina—Chapel Hill. A former mathematics classroom teacher in her native country of Iceland, Olof Bjorg Steinthorsdottir teaches mathematics education courses in the Elementary Education and Culture, Curriculum and Change programs. Her scholarly interests include the teaching and learning of mathematics among students in pre-kindergarten through middle school, specifically students' understanding of mathematics and how teachers can use that understanding to make instructional decisions. Steinthorsdottir also serves on the planning committee for the First School Initiative of the Frank Porter Graham Child Development Institute, a proposed school for three-year-olds through third graders. Her work addresses the curriculum and instruction to be implemented in the First School and ways to incorporate continual professional development into the format. Steinthorsdottir is active in the mathematics education profession locally, nationally and internationally. In collaboration with educators in the North Carolina Partnership in Mathematics and Science (NCPIMS), she developed and implemented professional development materials for elementary teachers that focus on the pedagogical and content perspectives of algebraic reasoning. She also serves as the U.S. national coordinator of the International Organization of Women in Mathematics Education (IOWME), an affiliate of the International Commission on Mathematical Instruction (ICMI).

Paola Valero is Associate Professor in the Department of Education, Learning and Philosophy, Aalborg University. Her initial background is in Linguistics and Political Science. Since 1990 she has been doing research in the area of mathematics education, with particular emphasis on the political dimension of mathematics teaching and learning, and of mathematics teacher education. Her research integrates sociological and political analysis of mathematics in different institutional settings, and different aspects of mathematical learning and teaching. She has published several papers in books, journals and conferences proceedings.

Robyn Zevenbegen is Professor of Education at Griffith University in Australia. She has worked across primary, secondary, post-compulsory and VET sectors of mathematics education as well as workplace settings. Her work in mathematics focuses on issues of equity and access, particularly for low SES, rural/remote and Indigenous students. She is concerned with curriculum, pedagogy and assessment as they relate to equity.